CAMBRIDGE LIBRARY COLLECTION
Books of enduring scholarly value

Technology

The focus of this series is engineering, broadly construed. It covers technological innovation from a range of periods and cultures, but centres on the technological achievements of the industrial era in the West, particularly in the nineteenth century, as understood by their contemporaries. Infrastructure is one major focus, covering the building of railways and canals, bridges and tunnels, land drainage, the laying of submarine cables, and the construction of docks and lighthouses. Other key topics include developments in industrial and manufacturing fields such as mining technology, the production of iron and steel, the use of steam power, and chemical processes such as photography and textile dyes.

A Treatise on the Practical Drainage of Land

Written in 1844 by Henry Hutchinson, this book, as the title suggests, focuses on the practical aspects of land drainage, advising readers to first consider the plan, cost, and mode of draining carefully. The treatise begins with a general address to the public which offers advice to landlords for dealing with tenant farmers, information on valuing land for fair rent, and ways of improving substandard soil. Lamenting that 'a great deal has been written by parties who really know nothing of the practical working of a system', Hutchinson, a land agent, valuer and 'professor of draining', writes from a zealous desire to educate the public correctly on the art of land drainage. Hutchinson's approach is scrupulously thorough, with separate chapters on shallow draining, deep draining, bastard draining, boring, and impediments to draining, as well as the history of land drainage in England.

Cambridge University Press has long been a pioneer in the reissuing of out-of-print titles from its own backlist, producing digital reprints of books that are still sought after by scholars and students but could not be reprinted economically using traditional technology. The Cambridge Library Collection extends this activity to a wider range of books which are still of importance to researchers and professionals, either for the source material they contain, or as landmarks in the history of their academic discipline.

Drawing from the world-renowned collections in the Cambridge University Library, and guided by the advice of experts in each subject area, Cambridge University Press is using state-of-the-art scanning machines in its own Printing House to capture the content of each book selected for inclusion. The files are processed to give a consistently clear, crisp image, and the books finished to the high quality standard for which the Press is recognised around the world. The latest print-on-demand technology ensures that the books will remain available indefinitely, and that orders for single or multiple copies can quickly be supplied.

The Cambridge Library Collection will bring back to life books of enduring scholarly value (including out-of-copyright works originally issued by other publishers) across a wide range of disciplines in the humanities and social sciences and in science and technology.

A Treatise on the Practical Drainage of Land

Henry Hutchinson

CAMBRIDGE UNIVERSITY PRESS

Cambridge, New York, Melbourne, Madrid, Cape Town, Singapore,
São Paolo, Delhi, Dubai, Tokyo, Mexico City

Published in the United States of America by Cambridge University Press, New York

www.cambridge.org
Information on this title: www.cambridge.org/9781108026642

© in this compilation Cambridge University Press 2011

This edition first published 1844
This digitally printed version 2011

ISBN 978-1-108-02664-2 Paperback

This book reproduces the text of the original edition. The content and language reflect the beliefs, practices and terminology of their time, and have not been updated.

Cambridge University Press wishes to make clear that the book, unless originally published by Cambridge, is not being republished by, in association or collaboration with, or with the endorsement or approval of, the original publisher or its successors in title.

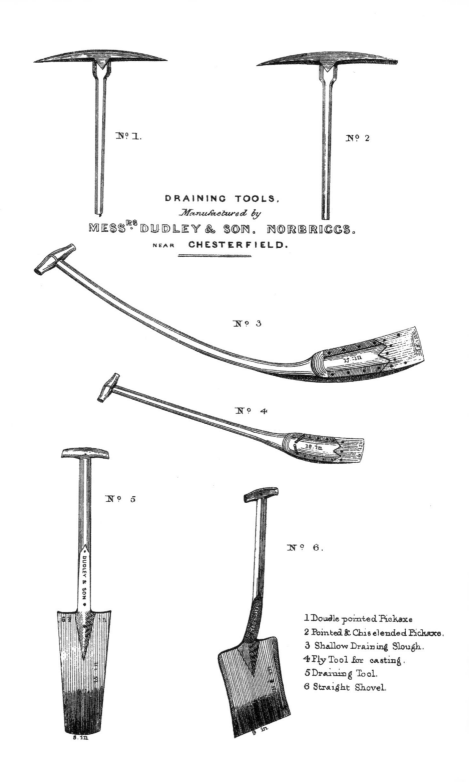

A TREATISE

ON THE

PRACTICAL DRAINAGE

OF

LAND.

BY

HENRY HUTCHINSON,

LAND-AGENT, VALUER, AND PROFESSOR OF DRAINING,

WALCOT, NEAR STAMFORD.

LONDON:
HOULSTON AND STONEMAN, 65, PATERNOSTER ROW.
STAMFORD: H. JOHNSON, ST. MARY'S-HILL.
AND ALL BOOKSELLERS.

1844.

TO

HIS ROYAL HIGHNESS PRINCE ALBERT,

OF SAXE COBOURG GOTHA, K.G., G.C.B.,

&c. &c.,

A GOVERNOR OF THE ROYAL AGRICULTURAL SOCIETY OF ENGLAND:

FROM THE INTEREST TAKEN BY YOUR ROYAL HIGHNESS IN ALL MATTERS CONNECTED WITH THE PROSPERITY OF THE COUNTRY, BUT MORE PARTICULARLY THOSE OF AGRICULTURE, AND WITH A DEEP SENSE OF GRATITUDE FOR THE ASSISTANCE GRACIOUSLY BESTOWED BY YOUR ROYAL HIGHNESS IN BRINGING FORTH THIS WORK, I BEG, WITH THE MOST PROFOUND RESPECT,

To Dedicate it to your Royal Highness,

AND HAVE THE HONOR TO SUBSCRIBE MYSELF,

YOUR ROYAL HIGHNESS'S MOST OBLIGED, OBEDIENT,

AND VERY HUMBLE SERVANT,

HENRY HUTCHINSON.

PREFACE.

For the last few years, scarcely any subject has been put forth connected with Agriculture, that has not had reference, in some way or other, to one of its first principles, *viz.* "Draining." Numberless have been the remarks in periodicals and other publications, as to the way in which it ought to be done, and many have written very powerfully upon the subject, not only endeavouring to secure attention on the part of the Agriculturalist to its importance, but also as to the mode of executing it. A great deal of useful information has been elicited; but the fundamental principles of Draining have not been clearly shewn by any writer, up to the present time: it remains for the Author, as a *Practical Drainer*, to lay before the public his plan, which he considers will secure to all parties the advantages of proper and effectual Drainage. It is not the Author's desire for literary fame that induces him to send his ideas into the world, but a wish to benefit the community at large.

The necessity for a Work of this description has long been apparent, and the Author trusts that the following pages will afford all the information required on this important subject.

The Author having had considerable experience in the different branches of his profession, will be happy to receive Communications from any Nobleman or Gentleman who may be pleased to honor him with their Commands.

STAMFORD, MARCH 25, 1844.

CONTENTS.

PAGE.

ADDRESS TO THE PUBLIC.................................. 1
Remarks addressed to drainers, landlords, land-agents, and farmers, on the value of land, and the rent to be put upon it.—The risk of property, and the position to be occupied between landlord and tenant.—The quantity of land to be drained in the kingdom, and the value of the produce after draining.—The investment of capital in the land, and the employment of additional labour with the money which we use in buying foreign corn.

INTRODUCTORY REMARKS.................................. 12
Thorough draining refuted.—The necessity for ascertaining what description of water land is subject to.—The rise and dip of water in the Eastern Counties.—Powers of Acts of Parliament insufficient to secure the outfall drainage of the country.—Water mills, and the obstructions they cause to effectual drainage.—Reasons for their removal, and other means shown for supplying their place.—The irrigation of land, and the necessity for upholding the water for supplying parishes in cases of drought.

ORIGIN OF DRAINAGE 39
Land drainage, its rise and progress.—Remarks upon drainers.

SHALLOW DRAINING.................................. 55
Its object.—The different methods pursued in various parts of the country for accomplishing it.—Thorn and kid draining.—Turf draining by sods, and turf draining from moss or turves.—Remarks upon by T. R. W. Ffrance, Esq.—Stone draining in various ways.—Draining by the mole plough or horse draining machine.—Gravel draining and various other modes reviewed.—Tile draining improperly done, and its durability.

SHALLOW DRAINING *(continued)* 88
Proper outfall to be obtained at starting.—Stone mouths for outfalls.—Drains to be set out at given distances.—Depth of drains.—Tools necessary.—Trial of drains, tiles, tile pieces, and flats.—Method of putting in.—Twenty inches not always to be the average depth.—Drains to be set out wider in open soils.—Draining of the levels and fens in the kingdom.—Work to be done by measure.—Prejudices of workmen in executing it.

DEEP DRAINING 107
Should be executed by the landlord.—Mode of detecting deep water.—Trial holes to be made before commencing the work.—Reasons why practical working men should be employed.—The level of the water to be taken at starting, and continued to the end.—Tiles,

how secured.—Planking and stretching the drain, to secure the sides from falling.—What to use for this purpose.—Where to finish the drain.—Branches into the main drain occasionally required.—Size of tiles to be used.—Where flat tiles cannot be procured, blue slate may be used.—Carrying outfall over dry ground.—Drain emptying under water into a dyke or river.—Reference to one executed.—Deep drains executed, with remarks upon, in Nottinghamshire, Lincolnshire, Derbyshire, and Northamptonshire.—Necessity for using the boring rods in deep drains.—Variations of cuttings, and quantities of water discharged from drains.

BASTARD DRAINING 143

Where required.—Piping, how performed.—Drain executed in Nottinghamshire.—Description of pipe tiles, with their use and application.—Remarks as to their use in shallow draining.—And the time shown when first used.

IMPEDIMENTS TO DRAINING...... 155

Drains are liable to be stopped up.—Stoppage of a deep drain at Welbeck gardens, Nottinghamshire, by horse radish roots.—The like in Edwinstowe meadows, by the roots of gorse.—The like at Saucethorpe, Lincolnshire, by the roots of an elm.—The like at Thoresby, Nottinghamshire, by the roots of a willow.—The process of scouring a drain.—The stoppage of other drains, and the causes.—Stoppage of a drain by the roots of an oak.—Water will shale in drains, with hints for prevention.

BORING 164

Its origin.—Artesian wells connected with it.—Boring for, in various places in France, and the result.—Description of boring rods.—Bore holes in Nottinghamshire.—Wells dried by boring.—Bore holes in Northamptonshire—General remarks upon.—And the value of tanks to farms for taking the rain water.

LABOUR 182

Expenses of.—How regulated in shallow draining.—Deep draining.—Bastard draining and piping, and also boring.

DRAINING TILES 190

Have been improperly made, and badly burnt.—The difference in value regulated by the carriage and price of coals, &c.—The Tweeddale Company's tiles and home-made tiles compared.—The sizes of tiles requisite for all descriptions of draining.—The prices paid for making and burning them.—The selling prices when made.—Gateway tiles, their use.—And remarks to those who may wish to have yards upon their own estates.

CONCLUSION 203

Remarks upon open dykes, &c.—The drainage of towns.—And the means afforded by the Act of Parliament for getting the drainage of estates executed.

ADDRESS TO THE PUBLIC.

REMARKS ADDRESSED TO DRAINERS, LANDLORDS, LAND-AGENTS, AND FARMERS, ON THE VALUE OF LAND AND THE RENT TO BE PUT UPON IT.—THE RISK OF PROPERTY, AND THE POSITION TO BE OCCUPIED BETWEEN LANDLORD AND TENANT.—THE QUANTITY OF LAND TO BE DRAINED IN THE KINGDOM, AND THE VALUE OF THE PRODUCE AFTER DRAINING.—THE INVESTMENT OF CAPITAL IN THE LAND, AND THE EMPLOYMENT OF ADDITIONAL LABOUR WITH THE MONEY WHICH WE USE IN BUYING FOREIGN CORN.

IN commencing this work, the Author is aware that a great deal has already been written upon the subject by men well qualified to judge of the merits and value of Draining, and a great deal has also been written by parties who really know nothing of the practical working of a system, which has for its object the gradual improvement of land of the very worst description, and making it become some of the best land in the country.

To those who have long studied and made themselves masters of the profession, the Author trusts they will kindly look over any imperfection that may arise in the statements laid before them; and to those who have written, without making themselves what they profess to be, *viz.*, "*Drainers of Land*," he hopes that he may be able to lay before them such facts and practical experiments, or results, as may lead them to the conclusion, that draining does not exist in Theory but in Practice. Any one who sees land in a wet state will naturally conclude that it requires draining, but the way to do it is the most important consideration—comprising the plan, the mode, and the expense. The writer of this work having given his close attention to it for many years, and having had practical working men under him who have understood their work, he has been able both to see it done for others and also to give his own directions, and to have the superintendance of, and the setting out of drains, and he now begs to lay before the agricultural world the result of his experience.

If he does not, in the succeeding pages, give grammatical statements of the various modes, ways, and carrying out of the works, the practical Agriculturist must overlook any blunders. The object of the Author is to make himself as plain and distinct in his statements as he possibly can, so that he may be understood by those to whom it is addressed, and that the benefits and advantages which are to be derived from following the plans laid down, may be such as to justify him in sending these, his humble ideas, into the world. And to those who may consider the following statements as forming a correct judgment of the mode of draining land, he will be ready to offer his professional assistance, either in taking the whole off their hands and draining it, or of attending and setting it out for them, and placing practical men on the work, who will not only do what they are set to do rightly, but will also see that those who are employed with them, will do so likewise.

Remarks upon landlords, as to the interest of themselves and their tenants, he trusts will be received by them with that kindness and forbearance with which they are offered, or intended to be applicable. To practical Agents brought up to a knowledge of their business, (the Author being an agent himself,) his remarks are intended to apply as a spur to them to look after the interests of their employers, and by a judicious application of the means, and timely attention to the interest of their employers, to prevent materials being used where they are not necessary, and by setting out and superintending the work, (as occasion will admit,) upon the farms which require draining, greatly benefit their employers, and not unfrequently save the tenants from ruin.

To those Agents who, by good fortune, acquire the management of estates, and who have had no experience but what they borrow from their friends, the Author begs them to get practical men to set out and superintend the work, and, in the face of the present times, assist the tenant in doing that which he cannot do himself, and by so doing enable him to procure his rent at the time it becomes due, instead of whipping him up, and threatening him with distress the moment he gets into arrear—and thus meet him with a face all smiles and thankfulness, instead of one as long as a fiddle, and grievances to recount sufficient to fill a sheet of foolscap paper. There is no class of men upon earth so grateful as the British farmer. I do not mean the Gentleman farmer, but the man who gets his living for himself and his family from his farm, and whose *all* is invested in it; and there is no man whose heart is so light and happy as the farmer, who, with his half-year's rent in his pocket, can meet his landlord, or the agent of his landlord, with a smiling face, pay his rent, enjoy his dinner and a pipe afterwards, and then return home with the pleasing reflection to his family, that he met his landlord and that he behaved kindly to him, and promised him various things which required doing upon his farm. It is this class of farmers too who want assisting, as when the harvest is over he has to begin to sell his spring corn to pay his servants' wages, his wheat, part to pay his Michaelmas rent, and before Lady-day his stackyard is cleared; great rises in the market he never can reap the benefit of, because he cannot keep a wheat stack to look at: it is only the *gentleman farmer* who can derive any benefit from the sudden advance of the market. The system, too, of giving back 10 or 15 per cent., and in some instances 20 per per cent., is bad in principle as well as in practice—*it gives the landlord a fictitious rental, and returns to the tenant a sum of money in bad times which he cannot afford to lay out upon his land, but uses for other purposes.*

The first step with the landlord should be, either by his agent or a person well qualified for the purpose, to put all his tenants upon an equality—that is, both good and bad land at its fair value as to rent; and in selecting the party to fix the rental, he should have a man of sound judgment, one who would act up to the Golden Rule, and who has had practical experience to enable him to do this. It has been too much the fashion in valuing, to select men who have looked more to the landlord's interest than that of the tenant, and in so doing have done both an incalculable degree of injury; the landlord, by giving him more rent than the tenant could well pay, and the tenant, by ultimately ruining him, and compelling the landlord to reduce the rental afterwards, to get another tenant. Let the Valuer put such a rent upon it as he would give himself if he had to occupy it, and then there can be no cause of complaint, and by so doing he would serve the cause of both.

Having ascertained the rental of the whole, place so much apart in each year for carrying on the permanent improvement of the property, and, above all others, that of having it well and effectually drained. After that, the buildings that want repairing, and then any other alterations which may be found necessary. If this plan were followed—and there are many noblemen and gentlemen who do follow it—there would then be no occasion for these abatements, the landlord would be certain of his rental, and the tenant would not be troubling his landlord or agent incessantly with complaints. If the bad land is considerably improved, and the outlay great, and no notice is taken of it at the time of making the valuation for rent, why then it may be necessary to add 5 or perhaps 7 per cent. upon the outlay, as may be agreed upon at the time, and this it will bear. But I do not mean this per centage to apply to the buildings, as no man can take a farm without buildings upon it, unless he be a man that can occupy two or three farms, and does not want a house; but even then, life is uncertain, and if one can farm a place without a house, the next tenant may require one, and therefore this also is necessary; *and it very rarely happens that land is so well farmed as where the tenant occupier resides.*

It may be argued by some that the plan here proposed would, in very good times, put into the pockets of the farmers too much money; but in answer to this I would say, who spends his money so freely as the farmer when he has it, in improving the tillage of his land? and in the bad times, who puts any money into *his* pockets to help him to meet them. It is a rare thing to hear of a farmer dying *rich* by *farming* alone.

The risk will be found by all practical men to be equal, and we need only take the last fourteen years as a sample; as, if he has good years, he has likewise his bad ones. The risk is the same too as regards the soil, for when the season is good for clay farmers, it is invariably bad for sand farmers, and *vice versa*; true it is, there are middling years, when both do pretty well; but take the course of seasons, this reasoning will be found correct. I cannot in this Work proceed to point out more particularly the interest of landlord and tenant; and I have been drawn thus far beyond the limits in endeavouring to secure attention to my favorite theme, "*Draining,*" and to prevent the pursuit of a system which can be of no real good to either party, and to secure the investment of *Capital* in the land, which will amply repay both the landlord and tenant;—the one by securing his rent and improving his estate, the other by increasing the production of the soil, and also by this the means of procuring his rent; and in the present times of putting him in a position to meet the future.

Upon the best evidence which can be obtained, it is stated, that there are ten millions of acres, or nearly so, in the kingdom, which require Draining. What an enormous quantity of land this is! and what an immense number of *souls* might be kept upon the extra produce which this land is capable of producing, by Draining alone! Every practical agent or farmer who has seen land drained, knows what an incalculable degree of labour and expense is saved in the management and manure, and what an extraordinary increase takes place in the product of the soil; in addition to these, there are instances, not one or two, but numberless instances, where land, which never grew anything but rubbish, has brought magnificent crops of turnips, oats, and wheat, after it has been fallowed. There is also other land that has brought oats and wheat without a fallow at all. Again, there is land which has been partially drained, which has brought (after being effectually done) crops double in amount; and other land, which has been more managed, has increased in quantity one-third and one-fourth; but, generally speaking, more. If then, by the simple process of Draining, land can be improved, so as to produce one-third or one-half more, what a field is here for the investment of Capital—safe, secure, and certain, if rightly managed—that is, not overdone at first in the rent. And where can the landlord so profitably invest his capital as in the soil, of which he is the lord; and what so pleasant to his sight as to look upon a flourishing and numerous tenantry, all reposing confidence in him: not that he is to consider them merely as tenants, but should look upon them with the feelings of a father, who, with their children, have embarked their all, in the cultivation of the soil, and that their respect for him should be the same: the *one cannot live without the other;* and therefore the landlord should think as much of a good tenant, as the tenant should of a good landlord.

But let us look a little further into this statement of the quantity of land necessary to be improved and further managed by Draining. It is generally admitted that the Almighty did not create man that he should be pined or starved, and we have sufficient sense given us to see and learn from past observations and past events, that as the population of our country has increased, the supply for their wants has increased nearly in proportion; true it is, there are adverse seasons when we have not an average supply, and then we are compelled to buy of our neighbours on the Continent; but there cannot be a doubt, nor a shadow of doubt, that this *country is capable of growing all the corn necessary for our own consumption.* Such being the case, why not at once attempt to do this, and keep our money at home for other purposes.

Suppose we have to buy one million quarters

of corn for our own use which we are at present short of, why not let us have five hundred thousand acres of this land so drained and improved as to produce it; for if from two and a half quarters of wheat which it now grows, it may be made capable of growing four and a half and five quarters, here then is the quantity required for our use and consumption, and this money, instead of being transferred to other hands, may be employed in the extra labour requisite for doing this, *at home*, and so benefit the nation in two ways, and not only for a time, but a length of time, and what is left undone in Draining, so as still further to increase the production, *Science* with *Practice* will eventually accomplish. These are important facts for the consideration of landlords, and demand their serious attention, and every one who has the welfare of his country at heart, will readily join in effecting and completing these improvements as soon as he can.

In speaking of labour, I am perfectly aware that it is not possible to employ every labourer you meet, in the drainage of land; and that for a length of time, it must be executed by a class of men who know how to perform their work; but in the end, if this large quantity of land is to be drained, long before it is half completed, an increase to this class will be made, and as you employ a certain portion in one particular department, it will afford an opportunity for those who do not understand, or are incapable of doing it, being employed elsewhere, and thus the benefits of keeping our money, and employing labour at home, will in the end be realized, and the greater the demand for labour, the less will be the demand upon the Rates of parishes for assistance.

INTRODUCTORY REMARKS.

THOROUGH DRAINING REFUTED.—THE NECESSITY FOR ASCERTAINING WHAT DESCRIPTION OF WATER LAND IS SUBJECT TO.—THE RISE AND DIP OF WATER IN THE EASTERN COUNTIES.—POWERS OF ACTS OF PARLIAMENT INSUFFICIENT TO SECURE THE OUT-FALL DRAINAGE OF THE COUNTRY.—WATER MILLS, AND THE OBSTRUCTIONS THEY CAUSE TO EFFECTUAL DRAINAGE.—REASONS FOR THEIR REMOVAL, AND OTHER MEANS SHEWN FOR SUPPLYING THEIR PLACE.—THE IRRIGATION OF LAND, AND THE NECESSITY FOR UPHOLDING THE WATER FOR SUPPLYING PARISHES IN CASES OF DROUGHT.

THAT Draining is a subject of very great importance, is a fact now generally admitted, and next to the cultivation of the soil, it forms, in most cases, one of the first claims upon the attention of the agriculturist. It is not until within the last twenty or thirty years that those claims have been properly attended to, and in too many instances not at all. In proportion to the exertions which are made, or have already been used in improving the agriculture of the country, there are still vast and extensive improvements to be made in draining the land, to render it permanent and secure, and to bring it to that state of perfection and that freedom from water, which would render it at all times safe and productive. That land is still capable of great improvements in addition to draining, and that, to use the words of many very eminent agriculturalists, " Farming is yet in its infancy," is a truth which cannot be contradicted, and one to which general testimony is borne, as every succeeding year brings with it new methods of Management, new Seed, new Implements, and, above all other things, an increased desire on the part of the agriculturalist to improve, perfect, and make himself acquainted with the improvements that are going forward in other districts, a thirst for knowledge, a spirit of emulation, and an anxious wish not only to benefit himself, but also his fellow man, that the productions of the age, or rather the earth, may progress in proportion to the extent of the increased demand of a daily increasing population. But amidst all this rage for improvement, if there is one thing more backward than another, and which is important to be attended to by farmers, it is the Drainage of the Land, as it is clearly shewn by all practical agriculturalists, that it is idle to farm bad, or strong land, unless you first get off the water, and having got it once clear off, you may do what you please with it, as it is not like man— it is grateful (if the term gratitude will apply) to increased production.

There is no department which I know of in

agriculture on which more money has been unprofitably expended, than in that of draining land. It has been thought so easy, that during the last thirty years, hundreds have tried it, thousands have been expended upon it, thousands of tiles have been unprofitably used, and improperly applied, by being put into the land at all depths, and without any regard to the cause for draining or the effect to be produced by putting them in. In addition to Tiles, *Stones, Turf, Turves or Peat, Thorns, Gravel,* and a variety of schemes have been tried, some of which have answered the purpose, but by far the greater part, in many places, instead of being put into the land to drain it, ought to have been used and applied in repairing the roads to get to it; and doing this, would have been more beneficial to the owner and occupier, than to have been placed in the land for the purpose of drainage. How many tenants are there, too, who have had materials given to them, who have known no more of the way in which their land should be drained, than the Man in the Moon; admitting even that they have commenced Draining in a proper direction, they have not given themselves the trouble when materials have been put in, to see whether the water would run off, or not; and consequently have had no level at all to guide them, and have placed them in, hoping that as the land was cut, it would run off the water, as a matter of course.

The first and most important step, in proceeding to Drain Land, is to ascertain whether it is affected and injured by Top Water, or that which falls upon the land; or whether it is injured by deep or Spring Water rising up in open porous measures of the earth to the surface of the land, and then inundating it to a considerable extent, in too many instances making it absolutely fit for nothing, or, in others, rendering the herbage upon it unsound and destructive to cattle, by causing rottenness in sheep, and other disorders in beasts, brought on from eating the rubbish grown upon it; or at other times so wet and boggy in places, as to be dangerous (exclusive of other causes) to turn cattle upon it at all, or even to get upon it with any beast of draught; and this is the land which, when once well and effectually drained, proves to be generally the very best land upon the farm, and the most productive. But these remarks will not apply to all land that requires draining; there are in the country immense tracts of land that require very little draining at all;—and again, there are likewise large tracts of land which only require, from their level nature, to be shallow drained, or rather surface drained. A great deal has, within the last few years, been said upon this species of draining, and there have been some bold enough to assert, that one description of draining will do upon all descriptions and qualities of land.

This I shall be bold enough to deny, and, at the same time, beg of the advocates of such measures to bear with me, until I have shewn, by *example*, that this is *Theory* and not *Practice*. Those who advocate this system of draining have given it the term of "Thorough Draining," instead of calling it by the simple term from which it derives its name: *viz.*, the furrow, and is to all intents and purposes "*Furrow Draining.*" I am not surprised at this error in the term applied, for it is no unusual thing for farmers, and more particularly farmers' servants, in the various counties of Nottingham, York, Lincoln, Northampton, &c., to designate the furrow, a "thurrow, or thorough," shallow draining being generally set out in parallel lines, and the last furrow or scouring furrow, being upon the top of the tile or drain laid.

When you have discovered what kind of water you have to contend with, or your land is subject to, the next is the mode of setting it out, and the practical execution of it when it is so set out; and although it has been asserted by many clever men, that none but a practical person can do this, nor he without a *Level*, I have seen a great deal done of all descriptions *without a level at all*, having no other and none other required but the level of the water. And I would remark here, it very seldom happens, unless it is for the purpose of ascertaining whether any and what fall has, or may be lost, in carrying the drain into execution, but what the eye of a practical and experienced man will detect; but in this I shall hereafter shew a great deal depends, and will depend, upon the abilities of the man executing the work, and who must have a knowledge of what he is going to do before he commences the work. There are a variety of causes assigned for the rise and dip of water from springs, and it is really extraordinary to find it in some places rising to considerable elevations above the level of the land-drains or the water-courses adjoining, whilst in others it lays below the level of those waters, and yet makes its appearance upon the surface of the land adjoining. In strictly investigating the rise of the water, I have, from all the information I have been able to collect, and also from my own observations made in cutting drains, found, that *all water on this side of the kingdom dips from the west.* When stating this side of the kingdom, I mean from the range of hills which run from the north through Yorkshire, Derbyshire, &c., by some called the "Backbone of England," and that all spring water has its source or supply from the west. I cannot undertake to make this assertion for the other side of the range of hills, but this may afford a subject for a more learned and scientific individual than myself, and determine whether the other portion

c

of the kingdom receives its supply from the same source; or whether the course of the underground streams or supplies are regulated by the flow of water from the rivers into the sea; or whether the whole is not regulated, in this kingdom, by supplies from the west.

It may be asked, how is it where spring water is shewn to break out on the *east* side of a hill, and flow back from the east to the west? It is possible that the water travelling under a valley, may be so pent up, and confined under the surface, that it must travel a considerable distance before it can find an outbreak; and then such outbreak may be on the eastern side of a hill; but this does not destroy the statement that such water proceeded originally from the west And again, on the application of this question: how is it, I may answer, that when the waters of the *Trent* are high, or rather there is a flood in the Trent, that in various parts of Lincolnshire, not one or two, but many, the flood in the Trent is ascertained, not by going to the Trent to see it, but from observations made long before the time of the oldest man living in that part of the country; these same indications exist at the present day, at a distance of miles from the Trent, taken in a direct line to the nearest point, and fully bear out the assertion, that water dips from the west.

Amongst the number of instances which may

be adduced and brought forward, is one within two miles of Lincoln Cathedral, about north-east of it, to the left of the road leading to Nettleham, on a farm lately belonging to Robert Wilson, Esq., but now of — Fytche, Esq., and in the occupation of Mr. Clarke, one of the leading Agriculturalists in that neighbourhood. The land of this farm appears to be at a very high elevation, in fact, to the eye, upon the same level as the ground upon which the Cathedral itself stands; and those who have ever travelled out of Nottinghamshire into Lincolnshire, or from the western parts of Lincolnshire, take the Cathedral to be built upon the highest ground in the neighbourhood. In this farm is a field which is pretty nearly covered with water in a Trent flood, and which disappears again as soon as the waters of the Trent subside. The water breaks out at the upper side of the field; an open ditch is cut part of the way into this field and an adjoining one, and the water gradually disappears again through the open measures of the earth, and again takes its underground passage elsewhere; there is no drainage to carry it off the farm, and therefore no chance of cutting direct into it and carrying it away, but you must wait the fall of the Trent. Another instance of this description occurs at Ancaster (the parish where the former dukes derived their title or which gave the name), in a field there, and it cannot, to arrive at this place, travel underground much less than thirty miles in a straight line from the Trent, in any way; and when it is recollected, that if taken as the crow flies across the Cliff of Lincolnshire, it is extraordinary that it should find its hidden passage to this distant spot before it breaks out. The field is covered over, as in the last case, and those residing in the neighbourhood are aware of the cause. It is quite possible it may have other outbreaks before reaching here, but this is the only one noticed as being remarkable.

Another instance occurs upon Dunston Heath, or upon the Cliff line of Lincolnshire, on a farm belonging to or in the occupation of Mr. T. Dawson. The water breaks out in one particular spot where two hills join, or rather a continuation of the hill with a hollow way between; it flows down the valley to a bog (past a place called Bedlam Barn), in a broad sheet of water, when the flood is high in the Trent; and as the Trent subsides, it gradually decreases its run: and this too has been observed for years, so as to establish its authenticity.

There was formerly a prevailing opinion in Nottinghamshire, amongst some of the old-fashioned farmers, that all the water which broke out upon their lands in the shape of *spring water*, came from the Derbyshire Hills; and however frail their opinions some years back, past experience has proved that those opinions were not accidental, or erroneous as in the majority of the drains which have been cut, those who have completed them have invariably found the greatest body of water proceeding from the west; and from observations which I made in the time of Tebbutt (whom I shall hereafter mention,) and since his time, it is more clearly shewn that this is the great fountain from whence it proceeds.

It is not necessary for this purpose, or more fully to bear out this statement, to pass into other countries for information; but in further corroboration of the rise and fall of the Trent, and its appearance at a distance, they will be borne out by reference to the works of various scientific individuals* (or even those who read the " Penny Magazine"), that both in France and Spain, and also various other countries on the Continent, there are waters which, after traversing large districts of land, again lose themselves in the bowels of the earth, and again take vent in other parts of the country or kingdom; amongst these may be mentioned the Meuse in France, and the Guadiana in the province of Estremadura in Spain.

These statements do little more than shew the hidden travels of water, and how far water will go, but do little as to the great question of

* Humboldt.

drainage or draining (as where these outbreaks take place no drainage is requisite); as in some cases, where nature or the formation of the earth has created an outlet, so in like manner she has also provided an inlet for the same, and others have made themselves natural water courses, so as to draw off the low waters, which form the main outlets for the drainage of the country.

There is another hidden supply which may be found in some places, where water upon a high level, and the land of an open porous nature at top, may admit of its entering into the ground, and yet not so as for it to travel to any considerable distance; and this water may be pent up as in a large basin, nearly on all sides, but with a few open cracks or fissures left here and there, and there it will gradually ooze out and destroy a considerable piece or tract of land; but these instances do not occur very often. I have seen water confined in this manner, or rather limited to its outlet, make its appearance on the sides of a hill, and within 200 yards of the same place again disappear into the earth, and take its rise again at a distance of a quarter of a mile, each time breaking out again at a fresh level of the ground; and thus in its progress to its final outfall, destroy or injure two lots of land, whilst that between the inlet and outbreak has been perfectly dry. In cases of this sort, one deep drain to cut off the first supply, and carry it off either by a natural or an artificial outlet, would have the effect of drying the whole; but if it should happen between the first outbreak and the second, that the land belongs to two different proprietors, the man who belongs to the top water may not be disposed to drain at all; and the bottom man must drain if he means to have his land dry, in order to secure himself; whereas if both had joined in one outlay, the whole of the expence of cutting one drain would be saved, and both alike benefitted.

This remark arises here in consequence of the nature or extraordinary rise of the water to be got off, and others will arise, shewing the value and necessity for the drainage of land being carried on with all possible care from the outbreaks, as not only much time and trouble may be saved, but also an incalculable degree of expence, by parties being united, where it is to their interest; and overlooking and setting aside all minor differences of opinion, join cordially and freely in this great work, and having accomplished it, *time* the great *end* and arbiter of all things, will shew them that what could not be brought about to their satisfaction at first, will be accomplished in the end. The further remarks upon these freaks of nature will arise in the observations upon "Deep Draining."

It happens that in this part of the country* we have none of those large bogs to drain which occur in other parts, and great portions of those which do occur, are drained by artificial means, or carrying off the water at a higher level by means of water and steam mills. *Where the use of machinery is not required, or rather the inclination of the ground is such that a natural fall may be obtained, there is no land either boggy or approaching to a bog, but which may be drained by deep draining and boring*—the first to cut off the supply, the last to bring up the underground water below the level of the drain, which cannot be attained or reached by the drain itself, and this will be shown more fully in speaking of boring. The greatest portion of the low wet lands in this district lie next the sea, and are not in any great measure affected by spring water — the principal portion of them being drowned by the water which falls upon them, and cannot be got rid of without artificial aid; or, if not injured by this means, are subject to the heavy inundation of land water from the hills, which must be got rid of in the same manner.

Land floods are often very troublesome to get rid of, and can only be effectually (upon parts so situated) removed by bringing an outfall from a very considerable distance to draw them off, or else by the aid before mentioned; but this sort

---

* Eastern Counties.

of drainage is either undertaken by extensive proprietors or bodies corporate, and belongs more to the Civil Engineer than the drainer of land: that is, both departments are necessary; but it does not follow that the man who can drain land by machinery upon low levels, should understand the laying of tile drains. One has been followed for numbers of years, the want of the other has only become apparent latterly, in consequence of the increased demand for the productions of the soil, and is a work to which few, if any, of this class have turned their attention. When, however, the road is opened and made passable, which before was considered impracticable, it is very easy for parties to travel upon it; and although one description of draining is for the purpose of drying the surface of the land, the other is equally necessary, so that the main outlets may not be choked up. It is one of the great evils of the present day, to be met with in numberless instances, that parties cannot make their drainage effectual, because the powers granted by the Legislature have not been sufficiently stringent; and although some great improvements have been effected in the last few years, a great deal more must be done, so that by one general law, (and without being put either to the expense of law proceedings, or to some tiresome, slow, and vexatious process,) parties may be compelled to make their drainage good, or else to levy a fine upon those so neglecting. If a man drains his land, and then cannot get off his water, what an annoyance it is to him; either he and his neighbour have differed, and he will not open his drain for him for an outfall, or else he is some cross-grained stupid fellow who adjoins, and will not let off the water, upon the plea that the dyke always did before his neighbour's time, and he cannot see why it should not do then. This may appear a very stupid reason, but I assure my readers it is one which is very often alleged, and it is high time such parties were made to see their *own* as well as their neighbour's interest. There are other arguments assigned, equally ridiculous and absurd, which it is not here necessary to enter into; the above will sufficiently answer my purpose.

The public are averse at all times to *Commissions* and *Commissioners;* as, generally speaking, a Commission carries on the face of it *a job*; nor do they at all like to be interfered with upon their *own domains*. I think it would, perhaps, be better, where the jurisdiction of the Court of Sewers does not extend, to appoint some person to take a district or part of a county, such parts being properly defined, whose duty it should be to inspect *all drains*, that is, all *main drains and sewers* from parishes, not including towns, as they are generally drained by a local act, (except such portions of parishes adjoining to towns, which have their outfalls in other directions than those connected with the town itself, or rather the drains running in from the outskirts of the town), that have other drains brought into them, or which form the principal outlet for the waters of such parishes ; tracing such drains from their outfall into the rivers, or other reservoirs for receiving the same : and in all cases where such inspector should find parties neglecting to cleanse and properly scour such drains, give them due notice to do so—and in case of non-compliance or default, to procure summonses, and bring them before a Magistrate, who should have the power of inflicting a fine upon them, not a heavy one ; and in case of second complaints, double the amount, and so on every succeeding offence. I do not recommend this commissioner or superintendant, I merely throw out the suggestion, as in all cases the acts of a responsible and paid servant are better performed than those of an irresponsible servant or agent ; and I think the appointment would ensure the work being more effectually done, as other parties, from good and kind motives, (though not the best for themselves,) are averse to getting summonses for their neighbours, and if they are compelled to get them, it leads to unpleasantness and causes ill-feeling ; not that this ought to be the case, but it is so, and always will be, whilst men differ in their tempers and dispositions. By placing this power in the hands of an individual, he alone would become responsible, and these petty differences in society avoided. The limited power placed in the hands of the Leet Juries is often very improperly, or hardly ever carried out to an extent which makes it of any real value; for even when parties are at fault, it very often happens that they are upon the jury themselves, and give some reason for their neglect in not completing their drains; or if absent, some one announces that they are about to do it, and should the powers of the court be carried out so far as to fine them, (that is, those neglecting), the amount put upon their neglect is so small, that the court is set at defiance, and it is unnecessary, by causing extra expense, to compel them to pay: and thus one of those old-fashioned customs, which might be of great service and value, is rendered incomplete by the working of the thing itself.

In addition to the various causes and impediments which so often meet our view in travelling across or about the country, there is one great enemy to effectual drainage seen in numbers of situations, and very rare indeed it happens that you see the land dry in the immediate vicinity of these obstructions—I mean *Water Mills*, or, to speak more properly, *Water Corn Mills*. In the present age we are told daily, that it is requisite to improve the land to the fullest possible extent, and that no steps should be left untried to make it (with the assistance of Providence) produce all it possibly can: the reasons for this increase are shewn most fully by experienced and practical men, and that the more we can increase the supply, the better it will be for ourselves and our country. To effect these improvements, most of the writers recommend *draining* as the first and most important step to take, and if this is to be done, all the barriers which exist or oppose its progress, must be removed, to ensure its being permanent and effectual. I have generally seen (unless on the line of a river, and even then in some instances) in the neighbourhood of a water mill, as much land injured as would pay all the rent of the mill (to say nothing of the immense cost which they are at within themselves, both to the owner and tenant), in some instances it would even pay the rent twice over; experience has, I think, already taught us that they are not wanted; that they may be done without, and that other means than the precarious and changeable ones of water mills, may be made available to supply the wants of the community with flour. And first, I will take the mill itself, with its supply of water. If a large quantity of rain falls it is too great, and the miller cannot grind, he has *too much water*, consequently the mill stands; if on the contrary a season of drought should come on, and last for a time, he cannot grind, he is short of water. In the summer he has too much and too little, from sudden rains and drought, and in the winter he has the same difficulties to contend with, from an over supply, or else ice comes down the stream, and again prevents him from pursuing his calling. It may be asked and naturally what remedy do you propose in lieu of this uncertain and unstable one? I will answer, the first and greatest improvement made in the present day, viz. *steam;* even the self-same sites may be made available and used for this purpose, the only addition being the engine-house, and the great source or principal supply being water; and if the coals should be brought forward and made use of as an argument against their adoption, I can only answer that most millers deliver their large supplies in the market towns, and their waggons may bring the coal as back carriage, and even if this should not answer the purpose, I would recommend some of them to travel into the neighbourhood of Sheffield, to Hackenthorpe, and there witness what man may accomplish by skill and perseverance, at the sickle manufactory of my friend Mr. Staniforth, who, because short of water, husbands his supply in reservoirs, and makes it available over and over again, and although living in a coal neighbourhood, yet makes a great portion of the firing necessary for his engine, from the turning lathe where the handles of his sickles are made; these, and such instances as these, teach man in a neighbourhood of scarcity, the means of husbanding his supplies and resources. Having thus shewn that the supply from this source is a very precarious one, and that other and more certain means may be adopted, for getting and ensuring this supply, I think there cannot be an argument for their continuance. It rarely happens but the occupiers of these mills are farmers, and rent their mills with the land; therefore, if as much land can be benefitted or improved by the removal of these mills, and as much additional rent gained from this improvement as would make up the landlord's share, no great hardship can arise by their removal; and the tenant, who is generally a farmer as well, if he can by the removal gain as much land as will pay the rent by the improved drainage, cannot be much worse, as it frequently happens when he takes to the mill he has got his business to learn, and will therefore have little difficulty in parting with it again should it be necessary; but if he adopts the plan of "going by steam," this will be unnecessary.

But the principal cause for their removal remains to be told or explained to the owner of property: it does not occur with their use or utility, or with the demand and supply from their work. Generally these mills have a dam head or reservoir, containing an extra supply in cases of drought (and to give additional power very often to the machinery at starting, as this they can controul by means of weir boards,) which reservoir is elevated to a height of ten and twelve feet above the level of the waters below, and in many instances the level of the water in the dam is as high as the chamber floors in the mill house, whilst on the side of this reservoir the land is still lower and falling rapidly away, and the whole has to be upheld by an embankment on the side of the reservoir. There are other instances where the water is held up by means of weir-boards placed at the height of a foot or more above the level of the ordinary site of the weir, and these are resorted to in cases of expected deficiency in the supply by reason of drought, and to give them an extra quantity.

One of the evils of this damming up of the water, is to cause it to flow back upon the land for a considerable distance and cover it over with water, whereas if it had its own free and uninterrupted course, it would contain very rarely more in the river or stream but what might pass along it without injury to any portion of the land, on either side of its banks. This is then a great and unnecessary evil, and may easily be remedied. Again; the natural course of streams is so often allowed to run in and out, and by constant washing upon either one side or the other of the banks down the valleys through which they flow, and make so many curves, turnings, and windings that the least thing or branch washed down with the stream, is deposited at these turnings, and by accumulating, form impediments to the free course of the water; that this of itself is a nuisance, without having the aid of a water mill or wier boards, to assist in these obstructions—the one, after accumulating for a time, will remove of itself, whereas the other cannot be removed without the aid of man. In all cases where these windings and turnings occur, I would recommend the owner or owners of the soil to straighten their course, even though it should be the division or boundary, of two parishes or a county; and in cases where the land does not belong to one owner but to several, to agree to a *give and take line*, so that it may be straightened and the impediments removed. It may be argued, that a great waste of soil may take place from the highlands by this course, from washings from rain, and that if water is allowed to run off too freely in summer time, in some parishes, great inconvenience would arise from allowing all the water to escape so rapidly. To this I would answer, that in all cases where it could be accomplished, I would recommend the formation of *Water Meadows*, but not unless the deposit carried in the waters would be likely to benefit the land, as the mere sending water over a lot of land, if cold, strong, and retentive, will not improve it, unless it is accompanied by sufficient mud, to cause a deposit to take place, and in this the parties must be guided by the locality of the district, and the neighbourhood from whence their supply is derived.

This rule will not apply to all lands, as, should the valley lands be of a hot dry nature, the sending a supply of water over them in dry weather, will have the effect of not only cooling them, but also of assisting the growth of the herbage upon such lands, and some very important results have taken place from adopting this plan upon land of the description here named.* I cannot here enter into the merits of Water Meadows, but must content myself with so much as comes within the scope of the present remarks upon water mills, and in making the works for irrigation, it will be necessary to put down such stop-gates or sluices, as to convey the water upon the high levels, and gradually float it over the soil intervening betwixt them and the usual water course, without offering any real impediment to the free run of the water in the river or stream, or of carrying back water upon the high lands. But if the land is so situated,

* The Duke of Portland's Water Meadows, at Mansfield, Woodhouse, Clipstone Park, Carburton, &c.; Earl Manvers's at Edwinstowe, Palethorpe, Nottinghamshire; all of them hot, dry, sand land, now useful Water Meadows.

and such obstructions or impediments arise, that it cannot be irrigated, either from the inequality of the land itself, or the hostility of parties to such measures being adopted, it may yet be found necessary to place some interruption in the stream, so as to leave the waters for given distances of one uniform height, and to ensure a supply in cases of drought.

Whatever plan is carried out in this respect, the *Natural* Drainge of the ditches running into it, and also the Draining of the land from springs or surface water, should be taken into consideration before such works are executed; due regard should also be had, that if the land adjoining the stream is porous, it may not be injured by soakage, as is not unfrequently the case on the margin or borders of rivers; but where this is impossible from the height of the stream, I would suggest that at a distance of a few feet from such water courses, the system of puddling might be adopted with good effect.

Having thus shewn a few of the evils arising from back water, &c., and a proposal for remedying the land adjoining, making it available either for irrigation, or for the better growth of the natural herbage upon it, I now come to another evil arising from the dam or reservoir, and its elevation above the level of the surface of part of the adjoining land. Along the hill sides in the neighbourhood of these mills (situate as they generally are between the hills), it frequently happens that a considerable number of outbreaks or springs of water make their appearance, and at various distances from each other, which flow down into the streams in the valleys, and whose courses are marked by a large quantity of rushes, not only injuring the land in their course to such streams, but searching into it where porous to a considerable distance. I do not attribute the whole of these to the water being upheld by such dams or reservoirs; but of one thing I am satisfied, that if they are not the cause of the whole, they are of a great portion of these evils; for if one mill is worked on one side, the other is worked on the opposite side, and when it is recollected that the side next the stream is always strongly embanked, to uphold and retain the water, it must be evident to every thinking practical man, that if the side or bottom of such reservoir is composed of open strata or porous filtrating strata, that the water so upheld will pass through into the earth, and travelling along confined or pent-up for a distance, will again make its appearance at the first outlet which offers, and from the elevation, it will have no lack of force (although its channel or course may be limited), to compel and urge it forward to its outbreak elsewhere; and as I have before shewn that water will travel a great distance under ground, it is quite possible that other lands, besides those in the immediate neighbourhood of the mills themselves, may be injured by upholding the water in this manner, and that to an extent scarcely conceivable. To remedy these outbreaks, no plan can be adopted but that of *Deep Draining;* and this, whether carried on the side of the hill in the line above the outbreaks, or by cutting directly forward perpendicularly into the hill, must be attended with considerable expense; and if, after the outlay, the object should not be obtained, this will be thrown away. I would seriously recommend to the attention of the owners of the land the removal, where practicable, of these obstructions to the efficient drainage of the land; and as many estates belong to the nobility and gentry, whose interest it is to benefit them as much as possible, the sooner it is done the better, and in those cases where the property of the mill belongs to an individual, it may be the interest of the owners of the land injured by it, to encumber their estates by a yearly payment, or else pay such sum by way of equivalent to the owner of the mill, as may succeed in its removal; but if, on the other hand, a new power is obtained in situations where it is desirable to continue them, no such expense will attach to this measure.

Note.—The Author is fully aware, that in making the above remarks on the injury committed by water mills, that he is endeavouring to establish a new era in drainage, and also machinery connected with water mills, and that a good deal of opprobrium and ill-feeling will be engendered, for a time, amongst some of the opulent and worthy owners of mills; time, however, will determine whether it is not for the interest of the community: to those friends to whom it is unpalatable, he can only ask for their serious and calm consideration of this important question.

ORIGIN OF DRAINAGE.

LAND DRAINAGE, ITS RISE AND PROGRESS.—REMARKS UPON DRAINERS.

THE origin of Land Drainage may be traced back for several centuries; and we are indebted for the open drainage of the country to the outfalls into the great rivers, or arms of the sea, or, properly to designate them, the main drains of large portions of our low lands, to the ingenuity, skill, and ability of our neighbours, the *Dutch*, as having been the contrivers and forerunners of the greatest amount of drainage which this country possesses. And amongst the great works to be handed down to posterity, as having added more to the drainage and improvement of the country in the early stages, those completed and set out by Sir CORNELIUS VERMEUYDEN will rank foremost, he having laid down plans, and also set out and executed, some of the best drains of which this part of the kingdom can boast, and which are allowed to be the most efficient drainage in the kingdom. These at the outset were absolutely necessary to free large tracts of land not only from the overflow of the rivers, but from the large quantities of land water brought from the high parts of the country upon the levels, and which overflowed immense quantities of land—having no sufficient outlet, or, in numbers of instances, none at all, and the land obliged to remain under water until the whole was absorbed by it, or else gradually run off with the absorption. From this first step towards drainage, thousands of acres have been rendered perfectly dry, free from land floods, and made useful, which, prior to drainage, grew nothing but reeds, sedges, rushes, and coarse grass. Since the time when these works were started, considerable progress has been made in *Civil Engineering*, and works left incomplete by Sir CORNELIUS and his successors have been carried into effect. And here it would be possible to enumerate, as having added to this department, some of the most eminent Engineers during the last and present century; but these are unnecessary in our present work; as, in addition to the drainage, the process of warping the land, by floating the tide waters over it in spring tides and returning it in the neap tides, was accomplished by several of these individuals, rendering land, before under water, and dry land of little value, not only available for agricultural purposes, but placing upon it a deposit of soil by settlement of these waters each time they were passed over it, that will remain available for the same purposes for generations; the deposit very often exceeding two, and from that to four feet in thickness, of the finest loam which can be produced, and calculated to grow almost anything, and this at a cost of from £8 to £10 per acre upon land worth from 5*s.* to 20*s.* per acre to rent, yielding to the landlord, where executed, a return of from £30 to £50 per cent. for his outlay.

Further improvements have been made in the drainage, by carrying out the main lines, and connecting others with them, so as to make the original drainage more complete. The uncertain assistance of the wind-mill for lifting the water from the different levels—or rather, *to carry the water of the high levels of the country above the lowlands* into the main outfalls—has been superseded by the aid of steam; and what in former days was a matter of great difficulty, is now rendered more certain and easy. Still greater improvements have, since it came into use, been effected, which fifty years ago the most sanguine could not even contemplate This drainage of the main outfalls, to a considerable extent, affects the great question of the surface drainage, as, without these aids, the water poured in from the lesser resources or supplies cannot be got rid of effectually. But although, since the first commencement, great progress has been and continues to be made in this branch of science and improvement, the progress of the surface drainage has been very slow. Numbers of practical and clever men have continued the above description of drainage since the time of Sir CORNELIUS, in addition to those quoted ; and amongst those who effected great and extensive inprovements, and to whom the country is in a great measure indebted for two of its principal canals, in Lincolnshire and Yorkshire, may be mentioned the late Col. ELLISON, of Sudbrooke Holme. To him may be attributed the Stainforth and Keadby canal, and also the Foss Dyke, as well as several other drains or soke dykes,—and who, by his skill as an engineer, laid the foundation of the great traffic existing between Yorkshire and Lincolnshire, through the medium of the Trent.

In addition to him, many others might be enumerated, but as this is not absolutely necessary, and only shews some of the first attempts at drainage of the main waters, it would be needless to apply for further information in this quarter. The various and vast canals cut and executed in the last century, some for drainage, and others for the transit of goods, would be many, and serve to fill a volume, if a sufficient meed of praise were awarded to the parties who have executed them ; and my only wish is to mention those which come under my own knowledge.

I shall commence with the first attempt which we have recorded of draining the land from the springs or surface water. There cannot, I think, be a doubt but that attempts had in some instances been made to dry land, but these would be chiefly by making open drains, or top gripping, as it was years ago, and is to this day still continued; and it is really a matter of surprise, and hardly to be credited, that after the lapse of nearly a century since its first discovery, or rather the discovery of its value to agriculture, that the attempts to effect the completion of it have been so few.

In the year 1763, Mr. ELKINGTON, the *Father of Draining*, was left by his father the possession of a farm called Princethorpe, in the parish of Sutton-upon-Dunsmoor, in the county of Warwick. The soil of the farm was very poor, and so extremely wet, that it had been the cause of rotting several hundreds of sheep, without at the time any apparent possibility of removing the evil, but threatening ultimately to ruin its proprietor. It appears that with this prospect before him, he began seriously to attempt to remove the cause of it, and to drain it, and which draining he commenced in the year 1764.

The field in which he first commenced was a wet clay soil, rendered almost a swamp, (and indeed in some places a shaking bog,) by the springs issuing from a bank or hill of sand and gravel, being the open meadow, where the water ascended and overflowed the whole of the land adjoining, with the exception of another gravelly piece on the hill side of the close, where the water disappeared, and again made its appearance lower down. In order to drain this field, Mr. Elkington cut a trench, about four or from that to five feet deep, a little below the upper side of the bog, where the water first began to make its appearance; and, after proceeding with it a considerable way under the hill side, he found it did not reach the main body of the water, from whence the evil proceeded. At this time, while he was considering what was to be done, one of his servants came accidentally to the field where the drain was making, with an iron crow or bar, similar to what is used for setting sheep trays or hurdles. It occurred to Mr. Elkington that this drain was not deep enough to answer the end desired, and being also desirous to know what the description of substrata was underneath, he took the iron bar from his man, and after forcing it down about four feet below the bottom of the drain, on pulling it up, to his surprise and astonishment, a great quantity of water burst up through the hole which he had made, and ran down the drain. This at once led him to the knowledge of water being confined much lower in the earth than what at that time was the usual depth for drains to be dug, and induced him afterwards to adopt what at that time was denominated the auger, as a proper instrument, (since better known as boring rods,) to apply in such cases. By this plan he not only accomplished the draining of this field, but by similar means the whole of the wet land on his farm. Thus did the discovery of *Deep Draining* originate in chance, the parent of many other useful arts and discoveries, alike beneficial to mankind; and truly fortunate it is for society when such accidents happen to persons who have sense and judgment to avail themselves of the hints thus fortuitously given.

The success of this experiment soon extended Elkington's fame in the knowledge of Deep Draining, from one part of the country to another, and after draining several farms in his own neighbourhood with equal success, he at last became generally employed in various parts of the kingdom; and indeed, at last it was almost impossible for him to execute half the numerous offers of employment that were made to him. In the latter part of his time he executed a great deal in the Midland Counties, and from his long practice and experience, he became successful in all his undertakings. He could also judge, from the appearance of the strata of the earth, where the water made its appearance, and where to take the course of springs that make their outbreak through the open measures of the earth.

A great many people were of opinion that

Elkington's skill lay in tapping the springs, without attaching any merit to his manner of conducting his drains, or carrying them across the land, and the execution of the working part of them; but the first notion of using the boring rods was accidental, and was afterwards the means of assisting him in his other works.

Thirty-one years after the commencement of *Deep Draining* by Elkington, and when his fame had become extended, we find the Board of Agriculture in this country entered into it very fully, and on the 10th June, 1795, a motion was made by the President in the House of Commons, that a sum of money should be granted to him as an inducement to discover his mode of Draining.

In the year 1796, Mr. Elkington's fame had reached the *Highland* Society in Scotland, when that intelligent body of noblemen and gentlemen despatched a Mr. Johnson, a Surveyor, to watch the progress of Elkington's works, and to report the result to them. After staying a considerable time, and examining the works as far as he was able, he returned to Scotland, and subsequently published his work, describing it as "Elkington's Plan of Draining;" it, like many other publications, contained many inaccuracies, which the experience of later years has tended to confirm.

Although Mr. Johnson appears in the character of Biographer to Mr. Elkington, it does not seem that, like the Prophet of old, he had the "Mantle cast upon him," nor yet that, during his continuance in the country, he had any portion of the ideas of Elkington himself conveyed to him for his instruction, the general outline of his work being chiefly from his own observations. Nor has it been the case with any Drainers to shew fully, or declare their ideas or reasons to the world. The death of Elkington shortly after this, put a stop to the further progress of Deep Draining, and left what little knowledge was to be obtained of this important work, in this country, to be gained from the pen of another countryman.

One-third of a century appears to have elapsed before any further great attempts were made in Draining, and which may perhaps in a great measure, be attributed to the unsettled state of the times; and after the conclusion of the *peace*, we may set it down to the want of funds, in consequence of the panics which followed the effects of war, and the sudden change in the value of agricultural produce. The land, in consequence of the high or war prices of grain, having risen 100 per cent. from its declared value in 1800, and dropping soon after the war, or within a few years, at least 50 per cent. of the former increase, this may in a great measure account for the stoppage of Draining; and it was not until rents became more stable, that any great progress appears to have been made, in urging forward this all-important and desirable object.

The next of whom mention may be made, as to his knowledge of Draining, is TEBBET. His works stand unrivalled as improvements, both as to execution and result, unless the works of the great Brindley may be placed in competition with him. His (Brindley's) extraordinary skill in the Bridgewater Canal, was a work which at the time was unequalled for the facility afforded in the transmission of goods from one place to another, which in the present march of improvement may, and will ultimately be superseded. Not so the work of which mention is here made: this will last for ages to come, as one of the greatest improvements of the soil ever attempted by man; *viz.*, the conversion of more than 300 acres of poor sand land, growing gorse, ling or heather, and a variety of other forest plants; the value of this land at 5*s.* per acre or even less at the time, (some of it being worth nothing,) is now worth at least £5 per acre, and forms one of the most beautiful lines of water meadow in the kingdom, or in the world. Too much praise cannot be awarded to the Duke of Portland, whose confidence in Tebbet, and whose knowledge and foresight, led him to form an estimate of the importance and value of this improvement in the land, by making, in the midst of a tract of dry forest land, a never-failing supply of green herbage for stock in the driest season possible, and which will be handed down to posterity as a lasting monument of what may be accomplished by skill and enterprise.

It was TEBBET, who executed the Duke of Portland's water meadows at Mansfield Woodhouse and Clipstone Park, Nottinghamshire. His knowledge of Draining was equal, if not superior, to that of Elkington, and was carried by him to an extraordinary extent, as the estates of the Duke of Portland, the Duke of Newcastle, and the Earl Manvers, will bear ample testimony, as well as the estates of many private gentlemen in the county of Nottingham; places which, before his time, were masses of bogs and quagmires, and covered with rushes and all descriptions of rubbish, (some of which had been drained, or attempted to be drained, with all sorts of things, and failed,) are now to be seen as some of the most useful, as well as ornamental pieces of grass land in the country. In the early part of his life he was a working man, and of extraordinary prowess, and it was really a sight worth going a distance to see the rapidity and strength displayed by him in removing a quantity of earth, or of performing any difficult part of the work in progress—and the anecdotes of his prowess and skill are numerous. In the execution of Draining, which was chiefly *deep*, no difficulty was too great for him to encounter, no obstacle but what he overcame, and no depth too deep to obtain the object in view, *viz.*, to *drain the land.* During his time, and since then, many have been the attempts of parties to drain land from *deep water;* and some very heavy failures have taken place, by Agents and others who have attempted to follow the steps of Elkington and Tebbet. I think the cause, if placed in the right quarter, may often be attributed to their being led by the men executing the work, or from having men to do it, who literally knew nothing at all about it. Several of Tebbet's men started to drain after being employed by him for some time, and each of these fancied himself qualified, either to set out work, or to execute it; and, by their being too hastily employed, threw away the money which really should have been used for draining the land;—and I could mention instances where more than from £1 to £200 have been expended on a single drain, and completely thrown away, by the employment of such men alone. I could recite numerous instances, too, where the parties who have had the direction of works of this description, have erred in like manner, but for the sake of themselves and their employers, I shall pass over them, and so prevent unpleasant reflections for both. There are many others who might be named as having accomplished some very useful drains, and amongst those coming immediately within my own knowledge may be mentioned the Messrs. Parkinson, Mr. John, of Leyfields, (late Agent to the Duke of Newcastle); and Mr. Richard, (Agent to the late Earl of Scarborough); these two have set out in various parts of the country some useful and important deep drains, which have benefitted large tracts of land; but I think if anything the palm may be given to Mr. Richard Parkinson, he having given the subject a much greater degree of attention than his brother, his engagements not occupying his time so fully. His farm at Napthorpe, Nottinghamshire, is a beautiful specimen of the benefits to be derived from Draining. It was here too, that he made the discovery that both descriptions of Draining are necessary to ensure the effectual drainage of land, and that occasions often occur where one without the other is of little use, or rather only useful to a certain extent. Other Draining might be mentioned, executed by both, not only upon land of their own, but also upon land occupied by them under others.

After the death of Tebbet, which occurred about twelve years ago, his sons succeeded him in carrying on the drainage and works for the Duke of Portland and others, but not with the same skill as the father; for although (unlike him) they had received a liberal education, it does not appear that they possessed the like ability

E 2

in draining, for, I believe, that up to the time of his death, he was silent to all as to the discovery of the art of Draining. The only one who imbibed any good and sound notion of Deep Draining, was SAMUEL SMITH, of Eakring, Nottinghamshire, (now living in the West of England); he undertook and executed some very useful drains for various Noblemen and Gentlemen; he also completed the line of water meadows from the termination of the Duke of Portland's Estate, at Clipstone, through Edwinstowe to Ollerton, and also drained the land and flats adjoining.

I could enumerate several others, in Draining, who have come within my own knowledge, but having attempted to shew the Origin, Rise, and Progress of Draining, and of some of those who have been engaged from the earliest period or commencement of *Deep Draining*, up to the present time, it will not be necessary for the purposes of facilitating the work, to give the names of more. Many of those who have got a tolerable knowledge of draining, have bought it at very dear markets, by having their work to do twice or three times over, and there are others who have been more fortunate, and secured it at a less cost; but generally speaking, and without casting any unwarrantable reflection upon my brother Agents, I would say, that it is more frequently got by them at the expense of their employers, and by scores never got at all. I will here mention an instance to shew how little progress this necessary assistant to good husbandry had made within the last few years, in numbers of places. The late Mr. J. S. Bayldon, Author of " Rents and Tillages, and Valuation for Poors' Rates," who went as Agent to Earl Manvers in 1833, and, as a Surveyor and Valuer, had a very extensive practice in Yorkshire and other counties, and also great experience in nearly every department of his profession, had never seen or known anything of Deep Draining or its effects, until he saw one (in company with myself) cutting upon his Lordship's Estate at Eakring, and inquired, upon viewing it, whether they were cutting a canal, so formidable did it appear to him, and who, for the first few months of his agency, had no very favourable opinion of the system, but afterwards became a convert to it, when he saw the practical results.

Thus far I have endeavoured, though in a limited manner, to give an outline of the Rise and Progress of Draining, and in my further remarks, shall endeavour to illustrate the subject with such practical and plain directions, as I am enabled to lay before my readers, and which, I trust, will be calculated to *instruct, and not to mislead*. I shall also mention the benefits and advantages which have been and continue to be derived from it; together with the operations and results: and to this end it will be necessary to divide each subject as much as possible, so as for it to retain its interest in the work; the subjects will comprise " *Shallow Draining,*" " *Deep Draining,*" " *Bastard Draining,*" and also the necessity for accomplishing and carrying out each by the assistance of " *Piping* and *Boring.*"

SHALLOW DRAINING.

ITS OBJECT.—THE DIFFERENT METHODS PURSUED IN VARIOUS PARTS OF THE COUNTRY FOR ACCOMPLISHING IT.—THORN AND KID DRAINING.—TURF DRAINING BY SODS, AND TURF DRAINING FROM MOSS OR TURVES.—REMARKS UPON, BY T. R. W. FFRANCE, ESQ.—STONE DRAINING IN VARIOUS WAYS.—DRAINING BY THE MOLE PLOUGH OR HORSE DRAINING MACHINERY.—GRAVEL DRAINING AND VARIOUS OTHER MODES REVIEWED.—TILE DRAINING IMPROPERLY DONE, AND ITS DURABILITY.

So much has been said and written on this subject, that it would almost be difficult to find any further information with which to illustrate it, although of so much importance to farming, were it not, (after all the statements which have been made,) as much excluded from the great portion of the community as when first started. Since the commencement of the system, it is really astonishing to consider the hundreds of acres that have been (or rather have been attempted to be) drained, and many too, by men of good understanding and practical experience in farming, which either have proved ineffectual, or have been so improperly done, as to require doing over again.

The object of *Shallow Draining*, and the way

in which it derives its name, is from thin cuttings or drains in strong retentive or close soil, which retains the moisture or rain from Heaven, and being impervious to wet in the sub-soil, by means of drains at given distances, is enabled to draw off this water, and leave it dry and capable of performing the law of nature, with regard to vegetation; and which this water, retained upon the surface, is calculated in a great measure to destroy.

In the first stages of Shallow Draining, the drains were neither laid down upon a system, nor was sufficient regard had to the duration or term which they were to remain, or that was requisite they should continue, to prevent being done over again; and I have often heard the observation from farmers, when speaking of Draining, that they have drained such a piece of land *twice* since they have been upon the farm, and the very appearance of that land in instances which I have known, has been such at the time, as to indicate that it required doing again. Now if this land had been done under the inspection of a proper person, and the draining (which ever sort it might require, for the probability is, that the way in which it was done would be as likely to dry the German Ocean, as to dry the land intended to be drained) properly set out, the first expense might have nearly covered the whole, and the course of cropping afterwards have realized full one-fourth more during each successive year, from the commencement of his tenantry upon the farm. I may be wrong in here making a statement, but it is one which I am sure will apply in too many instances, where draining has been done, and that is, that too little attention has been bestowed on this subject by the Agents of Noblemen and Gentlemen, in looking after the tenantry and making themselves acquainted with the best practical method of draining land; and conveying the same to the tenantry, so that the tenant might be saved an unwarrantable outlay, and the land improved for the benefit of the landlord, as well as the tenant; and it cannot but be disadvantageous to both, if the work has to be done over again.

It has been too much the practice to give materials for Draining, and to see nothing more of the work than the landlord's portion of the expence; but instead of this plan, of allowing a a portion of materials and labour, it would be much better for both that the whole of an estate, when about to be drained, should be put under the management of a proper person to set it out, and let him have practical working men under him to execute it; thus the labour and also the responsibility will rest with him. If the landlord is to do the whole, then let a per centage of £5. or £7. per cent. be put upon the outlay; but as Shallow Draining is always considered a *tenant's job,* let the labour be paid by the tenant, and the materials be found by the landlord, (still subject to the same inspection,) and the difference between them will be very trifling, when the carriage of the tiles or other materials is added to the labour of the work, or rather set against the materials.

I shall here enter into a statement to shew the various methods which have been pursued in different parts of the country and kingdom, and also endeavour to point out the mode, and the materials used for effecting the Shallow Drainage of the land, including those which I have seen myself, and a few others which have come under my knowledge; together with a little that I am recently indebted to a friend of mine for, on the subject of Gravel Draining, heretofore unheard of in this part of the kingdom.

The use of *Thorns* and also of long thin *Kids*, may be named as among some of the earliest attempts of draining: large quantities of this description of draining are yet to be seen in various parts of the country, particularly the Eastern Counties. The effects of this species of draining became apparent in a few years though some have been known to last from ten to twenty years, (not as effectual drainage, but have continued to run) but the majority of them, from decay, and in other instances from the introduction of vermin into the drains, very soon became filled up, and ultimately rendered useless. I think upon the strong clay, the decay has generally been less rapid than in open, porous soils; from what cause I am not prepared to state, but the destruction of the drains has also been more frequent from vermin in the open soils than in the strong retentive clay.

I had occasion to cut some shallow drains in a piece of strong clay, in the month of May 1843, and the men opened a quantity of these drains, but they were so completely made up, and the wood or sticks so rotten, (looking as if they had been wedged or rammed in,) that they were of no use at all in the land. I likewise saw a lot of drains of this description cut through in Thoresby Park, Nottinghamshire, about six years ago; these were put in to dry a flat of land, lying in the park, from twenty to thirty years before, but were of little or no use, and rendered completely so, long before that time, by Tebbet, who cut a deep *Tile* drain, and dried the whole of it; and this piece of land, from growing rushes, &c., as high as a man, and so full of hassocks as to be impossible to find the deer amongst them, is now one of the finest pieces of land in the Park; but, had the wood drains remained, and nothing else attempted, the probability is, the land would have been in the same state now. The reason for opening these drains was to ascertain the direction of one of Tebbet's drains.

Amongst the next attempts at effecting the Surface Draining of Land, may be mentioned *Turf Draining*. I have known instances of Turf Draining that have stood a great many years, and there are various parts of the country where it is yet practised to a very considerable extent, in places where tiles have not yet been made, and where stone cannot be procured, and if left to me to decide, I would rather use turf than stone in the present day; but I have no intention of recommending this description of Draining at all, except for purposes hereafter mentioned or explained. I am at the same time free to admit, that if a lot of Land can be got dry for a term of years, from 15 to 20, it really is important to those who have the means of getting turf, to consider whether they had better not have it done (due regard being had to the directions of the drains, the width and depth, and the nature of the soil,) than to wait until such times as they can procure tiles; but as the day is not far distant when tiles may be had in all parts of the kingdom, by the Marquis of Tweeddale's and other patents, I do hope that all Draining will be done with tiles, and done securely and effectually, and that no more *experimental Draining* may be tried, but that landlords, adopting the same plans, their estates may be perfectly and *thoroughly* drained; and the drainage of the land rendered permanent and secure.

There are objections to turf, but they are not numerous, and there are *two* descriptions of turf used in Draining; one is the turf or sod pared off the land with a paring spade of the size and thickness sufficient to lay two sods together, or rather rear them one against the other; and the other is the turf cut in the fens and mossy parts of the country, and called in some places *Turves*. These latter are a more stable body than the former, and are only known (except for the purposes of fuel) in the districts where they are dug; they can be cut to any size and when dry are like bricks in form, and are compact. Draining of this description I have not seen, but from "*The Magnet Newspaper*," in 1839, I saw an interesting statement of the mode of performing it, and I here give the remarks made upon it by T. R. Wilson Ffrance, Esq., of Rawcliffe Hall, at a meeting of the Garstang Farming Society. In quoting Sir James Graham's statement relative to Tile Draining, he states, "Sir " James Graham gave the cost of 70 roods " of Draining, at 30 inches in depth, which " was done at 4*d*. per rood. He agreed with Sir " James in giving the depth at 30 inches, and in " laying the drain 10 yards apart, but as to the " material to be used in Draining they differed. " Sir James recommended Tile Draining, the " cost of which for Draining and going through " with the subsoil plough was £6. 18*s*. 4*d*. per

" statute acre. He (Mr. Ffrance) would describe
" his way of draining to them. Instead of using
" tiles, he drained with turf, which cost him
" 2s. 9d. per thousand, *viz.*, 2s. for cutting, 3d.
" for whins to dry, and 6d. for stacking; he cut
" them a foot long, and 28 turfs would drain a
" rood of seven yards. If 1000 turfs cost him
" 2s. 9d. 28 would cost him 1d. He would now
" calculate the cost of Draining an acre of land
" or thereabouts, to avoid decimals. Suppose,
" he said, seventy yards, and he put the drains
" at ten yards distance, seven drains at 70 yards
" would give 490 yards of drainage, at 4d. per
" rood, but the material cost him 1d. per rood
" only, and there was an important feature in
" the expense, the cartage of the turf as com-
" pared to that of the tiles. At present let him
" recommend to the farmer *not to put back his*
" *clay* into the drain, but to cart it to the next
" Marl Pit. There was then the expense of
" cutting the drain, &c., 4d., of the turf 1d.,
" and of *carting* the turf to the field and the *clay*
" *from* the field 1d., making 6d. per rood, and 70
" sixpences gave £1. 15s. 0d."

" With respect to the formation of his drains,
" he might state that they were 11 inches in
" breadth at the top, and tapered off to the
" bottom to $2\frac{1}{2}$ inches. The turf was 4 inches
" and was cut to 6, but it dried up a little."

" The drains were 30 inches deep, namely,

" 6 inches of swallow, 4 inches depth of turf,
" and full 20 inches between the surface of the
" field and the top of the draining turf. Some
" persons adopted the shouldering in a turf
" drain, but he did not, and made it taper from
" 11 inches to $2\frac{1}{2}$ inches, a regular slope. The
" turf was put in dry, and when the wet came
" the turf extended, and squeezed itself into a
" firm position. A cart wheel would go over
" these without disturbing them, neither were
" any of the drains filled up as had been sup-
" posed from the narrowness of the drains at
" the bottom. The fact was $2\frac{1}{2}$ inches was quite
" wide enough, it compelled the water to run in
" a continuous stream, whereas, if spread over
" a greater surface, it would run in eddies, make
" deposits, and choke up the drains."

Thus far the above appears all very reasonable, and I have no doubt would, to a considerable extent, answer the end intended, and the cost of doing it very moderate; but suppose the subsoil of the field alluded to, instead of being strong clay, had nearly 18 inches of soil and clay on the top, and underneath this it had been white or red sand, hard or soft, would this description of Turf Draining have answered? Instead of the sides remaining solid they would in all probability have bulged in, and silted up, and then the value of the tiles over the turf would be more apparent; but this I shall shew more fully in describing Shallow Draining itself. Mr. Ffrance speaks of carting the clay into the next Marl Pit; why not spread the clay at the time it is dug; it is quite evident that all subsoil thrown out will incorporate with the surface or top soil, and it also undergoes a rapid change on being exposed either to heat or frost; and as this sort of Draining is generally performed in the winter, it would have a better chance of falling; in fact, it rarely happens but it will fall of itself, and mix or incorporate with the other, without being at much trouble in spreading it. It will often happen that the subsoil is a kind of marl, and will do a great deal of good after it is incorporated, in making the surface soil work lighter.

The next description of Draining is *Stone Draining*, and large quantities have been executed in various parts of the country, but more particularly in those districts where stone abounds: by some it is broken up and placed in small broken pieces for a given depth, in the drains, from six inches to a foot; by others, it is done by placing two stones to rear against each other, and then a few small ones placed on the top; and by others, with layers of stones by the sides of the drains, and then filled up on the top with a quantity of broken stone.

This, or rather some of these descriptions of Draining, are performed to a very considerable extent in some parts of Yorkshire and Derbyshire, and the southern parts of Lincolnshire, and in Northamptonshire and Rutlandshire ; beyond these counties I do not take cognizance of the actual working of the system, as to them my observations have been principally directed.

With this description of Draining I will now proceed to make my remarks. That Stone Draining will answer, and that for some years, I am ready to admit, and the place or land where it answers best will be upon strong heavy soils, with a strong retentive subsoil, and its greatest possible depth hitherto, when executed, has been from twenty to thirty inches, and in many, I may say scores of instances, the heavy soil has been thrown in again, thereby rendering it as completely fast and impervious to water as it was in the first instance.

Upon light soils, or with open porous subsoils, it is not of the slightest use, for not only are the drains liable to silt up, but they are exposed to the inroads of rats, moles, and other vermin, which, by getting up, let down the soil and sand, and ultimately stop them. I have opened, and seen opened, a great many upon both descriptions of land, and during the past year (1843) have drained one field (with two drains cut under my own direction), that had as many stone drains in it as would cover every inch of the land with stone if they were spread out. The drain alluded to is Bastard Drained, and marked No. 2.

This is not a solitary instance. Half a mile further I shallowed and deep drained a piece of land which had been stone drained in every furrow. I made all the shallow drains straight, and instead of having the drains at all distances and winding round the side of the hill, (which used to be a favorite mode of ploughing some years ago, and is yet continued in many places), I set the drains at eight yards apart, and for the scouring furrow to be upon the drain, and the field now presents the appearance it ought to have.

The Marquis of Tweeddale, in some published remarks on Draining a few years ago, (and he is no mean authority on this subject, having given it a great deal of attention,) says, "*The objection I have to the broken stones is, because they are most expensive, and neither kind of stone filters the water.*" This remark has reference to stone upon tile; therefore, if broken stone is of no use upon the top of the tiles, it is less likely to be of use in drains by itself.

In numbers of instances drains have been cut from eighteen to twenty-four inches, and I may add in some places deeper, even to thirty inches, and the stone filled in within six or eight inches of the top of the drain. Setting aside the enormous quantity of labour, in leading, breaking, and getting these stones, how very wrong it must appear to the parties who have effected this description of Draining, should they have any wish to subsoil their land; and this description of Draining has had no reference to the quality or staple of the land drained: it has been applied to the strongest and most retentive clay, and in other cases to the lightest porous sand. No regard has been had at all to the future, and consequently in a few years it will require doing over again, provided the tenant is a man *moving with the times*, having a liberal landlord, and an active and enterprising agent to look him up, and give him a lift when wanted; but if the occupier is one of the old school, and will not take to these *new fangled notions*, the probability is, (unless he is unfortunate enough to be turned out,) that it will last his life, satisfied in his own mind that once doing will be sufficient.

It very rarely happens that in the carting of the stone, (there are exceptions, and it is well there are,) that you find farmers calculating the cost of *team labour* in fetching the stones required for the purposes of Draining, and it really forms a very important item in the expense, if it is calculated according to the distance or the number of loads led in a day: but even if these are not taken as the basis on which to make a calculation, let them value a cart and horse and boy at 5s. per day, and a man with two horses and a cart at 10s. per day, or setting these two aside, let us take boys and men together, and allow 7s. 6d. for two-horse carts, and then try the cost per acre of leading only—to say nothing of baring the land and getting the stone, which generally amounts to 6d. per cube yard, where it is given; and then the stone has to be broken, unless it be shale stone, and used in rearing up in the drain, (instead of being broken and put in,) and the small stones put in at the top.

But there are some who will say we can get the stone without any labour in breaking, it being all ready as soon as taken out of the pit—granted; and here the labour will be less; it is possible in numbers of situations to get stone without being at any trouble in breaking; but if my advice on this subject (given after due consideration) is taken, they would *never be got at all*, except for the purpose of repairing the roads, where they would be really useful; and in too many instances, in roads to strong land farms, they are most required, for, with the exception of the public roads leading through villages, and which it is absolutely necessary for the parish to repair and keep in order, it is possible to find hundreds of instances, where the moment you diverge from such roads to get to a farm outside of the parish, you find it almost impassable, and therefore the utility of this recommendation would be far more apparent here, than in being used by pulling the *horses nearly to death* to get the stones on the land, and to be of no real use when they have got there. I should have no difficulty in reciting scores of instances of this sort, but they are unnecessary, and a man has only to cast his eyes around him in such localities as I speak of, and too many instances will be afforded of this description, and where, if you were to attempt to pass any remarks upon the occupier, you would be sure to meet with some such excuse as the Dutchman gave to the American's invitation to ride on the railroad instead of his own dirty conveyance and worse road—" Mein father travelled on this road, and I do the same,"

I shall give one statement here, (more is unnecessary,) which occurred in 1843, or rather, between 1842 and 1843, of the actual cost of stone draining a piece of land in the county of Northampton, and I shall leave the advocates for stone drains to draw their own conclusions. The quantity of land actually drained was about five acres, and the tenant, an enterprising man, did it himself, as he found that if he intended to have anything off it in the shape of a crop, it would be necessary to drain it, and as the use of tiles for draining was unknown to him, or rather he had never seen any done, he decided upon draining it with stone, and completed it. It was after its completion, that I had an opportunity of seeing him and conversing with him upon the subject, and I pointed out to him the value of tiles over the stone. After inquiring how many loads of stone it required, he stated that he had fetched one hundred and sixty loads for the purpose; and I enquired what he could lead them for per load: he thought he could lead them for 3s. 6d., and he should not like to do it for 3s. My impression was, that he could lead them for 2s. 6d., the distance not warranting a greater allowance; but the roads, particularly on his farm, to the field in question, were dreadfully bad. I will therefore compare the difference, allowing him for the leading, and the stones he paid for at 1s. per load at the pit. The account will stand thus :—

	£ s. d.		£ s. d.
160 two-horse loads of stone, at 1s.. ...	8 0 0	1,500 drain tiles, No. 3, at £3. 10s......	5 5 0
Leading, filling, &c., to the field, 2s. 6d.	20 0 0	5,000 No. 1, at 28s...	7 0 0
	28 0 0	Leading do. from the yard, a mile	1 15 0
Cost of tiles	14 0 0	Cost of tiles........	14 0 0
Difference	14 0 0		

This is one instance. I could name another equally wide in amount, in the same county; but it will be said that the *great* item is for leading the stones; I admit it, put them at 1s. per load, and even then the balance is in favour of the tiles; but this is not all the evil, I have said nothing for the labour. I have allowed that to be equal, but that will bear a comparison, as, looking only at the difference between a man putting tiles into the ground and putting in the stones, he may stop at ten or twelve acres of stones, whilst one of my men will lay forty acres of tiles, if the ground is ready to receive them. Now let us look at the time occupied in leading the stones to the draining: according to his own showing they rarely led more than four or five loads per day with his team, consisting of four horses; and it will be seen that it occupied those horses and two lads and a man for *five weeks*. Is this nothing when a man has employment for his horses upon his own land: they could not lead in wet weather, and when it was dry they ought to have been employed elsewhere. The tile yard was about a mile from his yard, and his men could have loaded the tiles in the afternoon and fetched them in the morning, without much trouble, and four or five loads would have seen them home; and this is not all, look at the excessive wear and tear to his harness and carriages, and the cutting up of the roads upon his farm, by leading one hundred and sixty loads, instead of five. I would then seriously recommend to the attention not only of landlords, but to agents, and particularly tenants, the above simple facts, and I do really think, that no man who studies his own interest will be tempted to drain any more land with stones. It does not matter what distance you fetch them from, even if you get them upon your own farms, they are not calculated for the purposes of efficient drainage;—and here it may be well to state, that some years ago in the parish of Beighton, in the county of Derby, the tenants of the Earl Manvers, preferred fetching tiles a distance of five miles, to drain their land, in preference to putting in the stone which could be got upon their own farms; and what man is there, who, if his landlord would give him tiles, would not sooner fetch *two* loads of tiles instead of *thirty* or perhaps *forty* loads of stone; and again, look at the carting saved upon the land alone, as the less you have upon it, the less trouble and expense it will be in cultivating afterwards.

It is not many months since I walked over a piece of land in the neighbourhood of Wansford, that was undergoing the operation of Shallow Draining by stones, and it will hardly be believed when I assert, that upon land having a fall of not less than three feet in a chain, they were cutting drains at a depth of about eighteen inches, and filling them up to within five or six inches of the surface with broken stones, and on the top of the stones was placed a quantity of stubble. If the land really required Shallow Draining, and had been ploughed up, that is, thrown up towards the ridge, and the furrow where the drains were scoured out very shallow, so that when the land was ploughed down again, it would have covered the drain to the depth of eight or ten inches, then there would have been some reason for having the drains so very shallow; but instead of this, the land was nearly level, and no plough could ever put more than four or at the farthest six inches of soil on the top of the stone, and should the owner or occupier plough this land deep, the plough, as a matter of course, would come into the drain. Subsoiling is out of the question, because, if attempted, the drains must come up again; and here then is a case where both labour and stone are thrown away. From what I could ascertain, the land was in the occupation of a large occupier, and he was doing it at his own expense; but here is one of those instances where *practical* agents ought to interfere, and prevent this unnecessary and ridiculous outlay; and it really, in cases of this sort, becomes a *duty* on the part of agents to do so, even at the risk of offending the parties, as it is not only injuring the estate, but, should the present occupier happen to die, it might pass into the hands of another and more enterprising man, to whom this would be a species of dilapidation, as in all human probability he would have to take the whole or a great portion of it up again. But this is not all connected with this draining; for, on examining the land, I found at the very top of the piece, where these attempts at draining terminated, the ground was full of *spring water*, breaking out in various parts upon the land; and that there might be nothing wanting, not even the eye of an experienced drainer to detect this, there was, half-way up the rising ground, and in the very centre of the land, a large pond, showing that the field was troubled with water from other sources, besides that which fell upon the land. Before any steps were taken to shallow drain this, it was necessary that a deep drain should be cut, to take off the supply of spring water, and then, if after a twelve month had expired, it had been found necessary to take off the surface waters, it might have been shallow drained, but not at the depth it is now done. This may be rather a digression from my statement in referring to the deep water, but as it comes under the head of stone draining, it follows it as a matter of course, both being connected, and both perhaps necessary upon the same land.

The next description of Draining brought into use within the last twenty years, and which is calculated to be of great service upon *one* description of land, is by the *mole plough*, or by some called the *horse-draining machine*. I should not be doing justice to it, had I omitted to notice it here, as I have seen very good effects from the use of it. The only land where this will answer is grass land, of a strong retentive nature, and here, I have no doubt, the benefits will be apparent for a number of years.

The mode of doing this is by means of a wheel strongly fixed by an iron anchor or cable into the ground, attached to which is a chain (varying in length from one to three chains long), and the plough is attached to this chain. The plough is of iron, the form of a mole, but considerably thicker and longer, with a straight iron forming a coulter fixed in to it, and attached to the top or drawing part where the chain is fixed. In commencing this work, it is necessary to have the outfall cut as for tile draining, and also in starting, at the end of each furrow, it is requisite to carry up two or three tiles, so as to secure the fall of the water into the main outfall, more particularly as the plough at each commencement of the furrow, does not give the drain sufficient draught, and it is also necessary to cut down for the purpose of letting it in at starting The wheel is then revolved by two horses, which, at first starting, is attended with some little difficulty, as they have each to pass over the chain, (the same as over the horizontal movement of a thrashing machine, except that it is higher), and it is not until they have passed several times, that it becomes familiar to them. Having performed the length of the chain, the wheel or gin is removed further, and so on in each furrow, until the whole is completed. It is also difficult to get your land furrows so straight as to secure its always running in the furrow, as it rarely happens that you see the furrows upon land laid with grass, set out straight, and consequently the plough must run in and out. The action of the plough through the clay leaves an opening of about two or two and half inches in diameter, and the coulter leaves an opening down to it about one inch; thus all the surface water is passed into the pipe or drain, and from thence into the outfall and from the field. It is necessary too, that the field should have a proper fall or inclination, as if the drains are upon a dead level they will not be of any use, merely taking it from the top and filling it up, instead of carrying it off. I think it will not be necessary to point out all the objections to this mode of draining upon arable land, as it must be evident to every practical man that the action of the plough over the opening made by the coulter, must ultimately fill it up, and then destroy the object which it was intended to perform. Upon grass land the results are different, as, with the exception of stock treading upon it, and the growth of the grasses over it, there is nothing likely to fill it up, the stock not injuring the bottom part, and the top only partially, and the grass not growing sufficiently close to prevent the water from getting into the bottom for many years. I before stated, that there was only one description of land upon which it would answer, and this grass of a strong retentive nature; its effects on arable land are very apparent, and unless the plough is carried through clay free from stones, it will be impeded in its progress, and the work inefficiently done, and if attempted through a porous or light loamy soil, it would have nothing to bind against, to form the drain and keep it open, and would consequently make up.

I saw upon a farm of a friend of mine in Nottinghamshire, a quantity of this description of draining performed, upon the quality of land I have described, and it answered exceedingly well. To a tenant occupying this sort of land, it is a quick mode of draining, and particularly where tiles are hard to procure; the expense too is not great, the work being done for about four pence per acre, of twenty-two yards, or in some places of twenty-eight yards, and I have known instances at even less. To those who wish to have the benefit of their land, instead of seeing it constantly saturated with water, where they can procure a plough, it is really desirable to have it done.*

* Since writing the foregoing remarks upon the Draining by the *Mole Plough*, I have tile-drained a grass field, in the Lordship of Little Casterton, Rutland, (the property of Mr. John Pollard, of Stamford,) which was drained about fifteen or from that to twenty years ago, with this description of plough, and in the furrows the line of plough was as distinct as on the day when it was done; but the line of

The next Draining to which I shall allude is *Gravel Draining*, or draining with gravel, instead of stone, and as I never saw it performed, or was in any district where it is practised, I can only refer to it as forming one amongst many schemes for draining land. The same objections apply to it, in its application to the purposes for which it is used, as to stone, with this difference, that whilst one has to be broken, the other has to be riddled. The carting and labour are alike. I shall notice it by giving a few extracts from letters of a friend of mine in Buckinghamshire, who has practised it, whose remarks are valuable as far as this description of draining is concerned, and who, I hope, will, like many others, become a convert to the advantages of tile draining, &c.

EXTRACT I.

" With regard to Draining, I have studied it
" a good deal and practised it steadily for the
" last two years, my ground requiring it much. I
" think it reduces itself to a very simple process

coulter was completely made up, and not one drop of the water falling upon the surface, could penetrate through the clay into the drains; clearly showing that even in grass land of the description and quality named herein, this Draining is limited to a given number of years, depending in a great measure upon the stock which it will carry. The appearance was exactly the same as though a mole had burrowed through the land.

' with me—my land is a compound of gravel
" and clay, with a rich loam here and there on
" the low parts of the fields and meadows. It is
" on a hill facing the south, with a large com-
" mon almost flat, and holding the water on the
" surface all the winter, above me; this body of
" water lying upon a soil composed of clay and
" sand, is eternally working its way, in obedience
" to the law of nature, down the hill, when it
" comes to a stratum of clay, it rises to the
" surface, flows over the clay, and sinks again
" into the gravel, and so again and again till it
" reaches my low meadows in the valley. I
" have good coarse gravel which I sift twice
" in dry weather; with this I fill my drains,
" varying from fourteen to eight inches thick,
" ramming them down well, which I consider a
" most important point, and covering with young
" furze, of which I have plenty. I prefer thirty
" inches for the common drains, and rather more
" for the main drains, and where I have clay, I
" fill them up to within six inches of the surface
" with the siftings of the gravel, which gravel is
" of a sandy nature. The drains which I have
" made have succeeded admirably; neither do I
" think that I have occasion to fear that the
" siftings will eventually work downwards and
" block the drains; the furze will survive till the
" ground is thoroughly settled, and the gravel
" being small and well rammed, will not easily

"admit it— add to this, a passage will have been thoroughly formed by the running of the water. In a few places where I found strong springs and slipping shingly gravel, I put in more stones or gravel, and covered with clay, putting some heath over the clay, to allow it being trodden flat, without adhering to the feet. The clay from the drains I draw to the nearest burning gravelly meadows, and this I shall continue to do to some extent before I dress them with manure. Thus you see my course is simple, and pointed out by the nature of the soil. I assist the clay by introducing veins of gravel, and assist the gravelly surface by a coat of clay. *I am aware that where the flow of water is great, my system is not perfect*, stones not letting the water pass rapidly enough; in that case, I am obliged to let my main drain remain open, but this is only in the valley where the flow of water is very great, and where nothing except perhaps a very large tile would carry it off, and then I don't know how the auxiliary drains could be united with the main one."

"My plan is expensive, but I think it is done for ever. I have not made any calculations as to the cost. I open the drains by the pole, varying in price according to soil, seasons, &c."

"I have not arrived at a decided conclusion with regard to subsoil ploughing, neither will I adopt it ever with clay: at present, I will

"wait to see whether the effect of draining the
"clay deeply is not enough of itself I am
"inclined to think that when the soil is loosened
"too much, the manure will be washed too
"rapidly through it, and that though some
"benefit is derived to the green crop (and in the
"first year a very great benefit), still I think
"that injury must be done to the wheat crop,
"and that the soil may easily be rendered too
"porous. I am seeking conviction, and will
"not make a trial till I am thoroughly satis-
"fied. *So much is written, I think, by mere
"theorists, that one has considerable labour in
"culling from the mass that which will bear the
"trial, without hazarding a failure.*"

EXTRACT II.

"The draining of clay is easy enough, but
"springs are the most difficult to contend with,
"and require, I fear, almost endless trouble. I
"am afraid I have only shifted the flow of water
"from some of my springs, and that, having
"brought it away from the higher parts of the
"fields, it has gone into the gravel again, and
"come up in fresh places; this, you will perhaps
"say, proves that my sifted gravel does not let
"the water pass freely enough, and that tiles
"would have been far better. I am inclined to
"think so myself—and I must say that your
"arguments, and a little more experience, will

" induce me to try tiles in springy ground the
" very first opportunity. I am not altogether
" persuaded as to the relative cost, provided it
" is necessary to fill up the drains with siftings
" or carted material, as in pure clay, the smallest
" tiles being £1. 15s. per thousand at the kiln,
" (five miles and a half off,) without feet or soles.
" My sifted gravel is nearly used, the remainder
" I shall appropriate to shallow draining in a
" clay meadow, and what more time I can devote
" to draining this year will be employed, I think,
" with tiles."

EXTRACT III.

" I have, as I intended, began laying in some
" drain tiles, having exhausted my stock of
" stones ;* and, as far as I can judge, I have
" reason to be very much pleased with the sys-
" tem—*and it is certainly a perfect holiday after
" the enormous trouble and delay of carting the
" stones, &c.* I am just now cutting through
" some soft blue clay in a low meadow, where I
" must use the sole under the tiles, the clay is
" so soft. Although I think I have sickened
" myself with Stone Draining, situated as I am
" without the opportunity of picking stones
" copiously, when they come clean to hand, and
" can be used as picked; still I cannot but
" believe that what I have done is done effec-

* Stones here mean gravel or boulders.

"tually. I think nothing can penetrate mine
"except the water; the small stones were put
"into the bottom, the men picking them out a
"little as they put them in from the cart, and
"the whole rammed down tight, with a good
"layer of furze over the stones. They are cer-
"tainly not equal to carrying off a heavy rain
"rapidly, when the outfall drains are made to
"carry too many auxiliaries. I have cut my
"roads about a good deal, but more particularly
"in leading the clay down to the gravelly mea-
"dows below, which I think a very good mea-
"sure, preparatory to bringing them into a good
"state.

"One bad feature is, that the stones, after
"being sifted and thoroughly cleaned in the
"summer and autumn, the worms work up into
"them, and soil them at the bottom for some
"distance—but I certainly think I have done
"with them. The point of durability remains
"to be proved; but I should think, being out of
"reach of the atmosphere, if even soft tiles are
"deposited and deeply covered, can hardly ever
"perish."*

The foregoing remarks tell their own tale, and need no comment scarcely of mine; they are the plain statements of one who has practised

* The extracts from the above letters have been made by the Author since writing the work itself, and are here inserted to carry out further his remarks.

upon his own land, and the correspondence extends over three months. If then, to a man of business, the conviction is brought home, that he is not only wasting his time, but also an immense quantity of labour unnecessarily, surely it will bring conviction to others, who have followed the same plans. There is but one point left unexplained, and that is, my friend's land is troubled with *deep water*, from his own statement, and he is Shallow Draining it, in order to dry it. He may, to a certain extent, get the water off, but all the shallow drains put together will never effectually remove deep water. He is not the first who has erred, by hundreds, in this way. It is impossible to deep drain land with anything—to do it effectually—but tiles. The facilities they give the drainer, in getting forward with his work, are superior to anything else that has yet been, or is likely to be, discovered; and the day, I trust, is not far distant when this important fact may become general throughout the kingdom.

In addition to the means resorted to already named, may be mentioned, Concrete Drains, Wood Drains, and also several others, which I shall pass over, merely applying the remark of Sir James Graham, in his communication to the Agricultural Society, "That skill in agriculture " does not so much consist in the discovery of " principles of universal application, as in the

" adaptation of acknowledged principles to *local*
" circumstances." But it may be important for
the parties who have used these means, to adopt
them in their vicinity for want of other means,
(*viz.*, tiles,) which would render their draining
more secure.

I shall have occasion to advert to pipe tiles
for Draining, in my remarks on "Piping," to
which purpose they are applicable. But amongst
the recent attempts at Draining, is one I cannot
pass over without a remark, which is, the use of
the draining tile inverted, or rather turned wrong
way upwards. The formation of the present tile
is calculated to set it either upon the land or upon
a flat tile steadily, and for the purpose of *resisting
pressure* on the surface; how, therefore, it is
possible for any one to do so by fixing the tile
the wrong way in the ground, I at present am at
a loss to conceive; and as to affording either
additional facility for the purpose of drainage, or
a readier mode of laying them in the ground, is
perfectly absurd and ridiculous.

About three years ago I examined a quantity
of strong land in Nottinghamshire, which had
been drained above twenty years before with tiles.
The drains were cut without reference to the fall
of the land, the line of furrow or ridge, and were
run about in all directions with outfalls, or rather
outlets, in three sides of the field. These were
cut, (at the time they were done,) by a man who,

because he had worked some few months under Tebbet, "the *Drainer*," was supposed to know all about it. I examined the outfalls, and, from neglect, there was not one but what was completely filled with rubbish at the mouth.

The tenant was then taking the drains up, lowering, and putting them in again, as he had found they were not of the slightest use; and it is scarcely to be wondered at, when the whole of the clay, which was very strong, had been placed in again on the top of the tile; and, from the time and necessary traffic upon the land, they were as completely wedged in as "pipe tiles," and would have carried water to any distance, but perfectly impervious as to letting any in from the top.

I persuaded him to take them all up, and lay the field out again; but as he was doing it at his own expense, and the landlord being unable to assist him, (short, too, of means,) I endeavoured to point out to him that it would ultimately repay him: his fears, however, were great, and he put them into the old places again, (without the clay,) and of course the land was considerably improved. The depth of some of these tiles did not exceed a foot, and that a great portion; the depth of soil about six and eight inches. A field adjoining was drained over ridge and furrow by the same party, and the outfalls made up in a similar way.

I carefully examined great numbers of these tiles, and found none broken or defective, and the whole of them were as good as on the day when they were first put into the land for the purpose of draining, although the depth was exceedingly shallow, and scarcely out of the reach of the plough. I have seen other drains, in addition to these, taken up and set out afresh, after being in the ground nearly thirty years, which have likewise proved to be good, and, had they been properly put in at first, would not have required moving—which clearly establishes their durability, even though improperly applied.

SHALLOW DRAINING (CONTINUED.)

PROPER OUTFALL TO BE OBTAINED AT STARTING.—STONE MOUTHS FOR OUTFALLS.—DRAINS TO BE SET OUT AT GIVEN DISTANCES.—DEPTH OF DRAINS.—TOOLS NECESSARY. —TRIAL OF DRAINS, TILES, TILE PIECES, AND FLATS, METHOD OF PUTTING IN.—TWENTY INCHES NOT ALWAYS TO BE THE AVERAGE DEPTH.—DRAINS TO BE SET OUT WIDER IN OPEN SOILS.—DRAINING OF THE LEVELS AND FENS IN THE KINGDOM.—WORK TO BE DONE BY MEASURE.—PREJUDICES OF WORKMEN IN EXECUTING IT.

THERE *is but one plan of Shallow Draining land*, or one way of getting off the surface water; and although years have elapsed since the first commencement of Draining, yet at this day there are hundreds with as little knowledge of this mode, or plan, as on the very first day when it was discovered that Draining was necessary, and essential to the improvement of the soil

To this plan it is my wish to draw the attention of Agriculturalists generally. And first, as to the field to be drained. It sometimes happens that land lays so advantageously, or of such a regular dip—or, more properly to designate it, I may call it *table* land—where it will only be necessary to have one outfall drain, that it requires but little trouble in setting out. But there are other lands, which either require the main outfalls to be carried to another level of the land, or where it may be necessary to have three outfalls, owing to the nature or dip of the ground. In all cases where land can be drained with one outfall, I recommend it, in preference to having more, as the greater the quantity of water through your main drain, the less likely it is either to silt up or admit of anything passing up it.

My remarks will now exclusively refer to Tile Draining, and I strongly recommend it in preference to all other means which can be adopted, as being that which is best adapted to all soils, and in all situations. Not that I would have parties stop where tiles cannot be procured; but to all those who can, and have the means of procuring them, they are decidedly preferable to anything else which may be adopted. Having stated my views with reference to tiles, I will now endeavour to lay down the mode of proceeding as adopted by myself, and which has been practised by numbers in Nottinghamshire and elsewhere, where this description of draining has been followed. Having, then, found the lowest point in the field for carrying off the water, the main drain, or outfall drain, should have its mouth there, at the side of the bank, and which mouth should, if possible, be a stone about twelve inches by nine inches, with a hole in the centre, chisselled through, about four inches in diameter, and with the stone cut to face the batter or slope of the bank, and the last tile laid so as to meet the stone at the opening. This stone will stand for years; for if the tiles are exposed at the mouth, they are apt to perish, and constantly require replacing.

Proceeding then from the mouth into the ditch, the outfall drain should be carried at a distance of not less than seven yards from the hedge, at a depth of not less than three feet on the average, and if the land rises in ridge and furrow, it may be necessary to have it six inches lower, so as to preserve the proper inclination to the end.

The outfall being cut, the next step is to set out the other drains in parallel lines, at a distance of eight yards, (not less); this distance will be found as far as is requisite, if the soil is strong and retentive. When I say parallel lines, I do not mean that you are to alter the appearance of the field, unless it be upon soil where you are warranted in so doing.

It sometimes happens that lands have been set out varying from seven to ten yards in width, and if these are tolerably straight in the furrow, I would not, merely for the sake of making them all at equal distances, bring them into eight-yard lands; but where it is important to get lands which have curved made straight, so as not to interfere with the future management of the soil,

I would so set them out, beginning with the straightest side of the field, and continuing at the distance you decide upon, at which the water will draw from the land throughout the field, and this will in clay be generally found to be eight yards; and where you cannot make them all draw into the main outfall, in parallel lines, lengthen and deepen one or more of the auxiliary drains, and by one or more crossings from the sides or ends, so preserve the uniform appearance in the drains or furrows throughout. The next point is the depth of the drains, and the mode of cutting them. It is stated that the depth should not be less than thirty inches, and in strong land, upon the old line of furrow, this will be more than sufficient; but if the drains are properly cut, eighteen or twenty inches will be found as deep as they need go. This I will endeavour to prove; and I may here state, that with the practical Agriculturalists in Nottinghamshire this depth is rarely if ever exceeded; but I mean the drain to average this depth throughout.

The next is the cutting by the workmen. Those whom I have seen employed, and are employed by me, are furnished with a hack or pickaxe, a straight shovel, a draining tool, and a slough.

The pickaxe can be made by any blacksmith, but a great deal depends upon its form, and the temper of the material, as some will wear out so much sooner than others, and are made so much handier to use. A great deal depends, too, upon the eye of the pickaxe, in its durability, as if the eye is welled instead of being punched or worked out in the solid, they invariably give way there, and the points should always be well steeled. These tools (that is, the shovel and the draining tools) are made by Messrs. Dudley and Son, of Norbriggs, near Chesterfield, and for form, shape, and durability, they are superior to any others of the kind manufactured. The draining tool is made fifteen inches long in the blade, steel tipt at the the corners, and is five inches across the bottom end.

The man cutting the drains proceeds with a line about three chains long, and sets them out straight; he then, with the shovel, takes off about four or five inches of the top, along by his line, and afterwards proceeds with the tool with which he, after he has once set in, and cutting a piece each time in the form of a V, takes out the whole depth with it, leaving the drain about nineteen or twenty inches deep ; but this is only done by the practical drainer, and you cannot expect every man, who takes a tool in his hand, can cut drains in this manner without having some knowledge of, or acquaintance with it.

The first six or eight inches of this may be soil, and the remainder of it strong clay, which he casts upon the ridge of the land as he throws it out, and if there is any soil attached to this, he cuts it off at the time of filling in. If the weather is wet, he will have an opportunity of trying the drains, and ascertaining if they will run—and if it is not, it will be better to wait for rain, or, if advisable to get on with it, a water cart should be procured, and taken upon the land, and every drain tried before a single tile is allowed to be put in; not that it will be necessary to keep them all open, as the work may be going on, and the tiles put in by degrees.

As soon as the drains are cut, and before the tiles are put in, the man who lays the tile takes his slough, (not an article made like a road scraper or hoe laid flat,) but a tool made of a piece of ash or willow tree, dished out in the receiving part to receive the crumbs, and shod at the sides and ends, to prevent wear and to cut clean—the first end of it being nearly flat, and about five or five and a half inches broad at the bottom, and about eight inches from the end, a foot or more of it is made hollow like a scoop; with this he proceeds to level the bottom, and to take out all the bits which may have fallen in, and to see that the water will run freely from end to end, and he at the same time ascertains if there are any soft places in the bottom where it will be necessary to use flats, or tile pieces, to rest the tiles upon, as it is not necessary to put them all the way along the drains, the bottom being sufficiently hard to do without. I have seen instances where the clay has become so soft, on being exposed to the atmosphere for a few days, and being rainy or wet at the time of cutting, that on laying in the tile it has sunk into the clay nearly over head. Here then it was necessary to put flats at the bottom of all the drains, or else, if put in without, the draining would be useless. In the soft places too, it will be necessary to take out a little more of the bottom, so as to preserve the uniform range of the tiles. The workman then proceeds to put in the tiles, not by laying them in just as they are placed upon the ground, but by placing one against the other, so that the tile when laid may have the same straight appearance as the drain which he has cut. Having placed in the tile, the best thing which can be put upon it is a layer of stubble; and if there should happen to be in the field a lot of banks, which can be pared, or sods of any description got, they may be put on the tile, or if there is no stubble, a lot of rough dry grass or refuse hay, after which the soil from the sides, and filled up. It is not necessary to put in any other rubbish, neither stones, boulders, brashy wood, or gorse. The drain once laid as stated, will draw the water from the land; the soil acting as a filterer, will always admit of its passing through, and *any other labour upon it is idle and useless.* The carriage of stone, which the Marquis of Tweeddale says will not filter; the putting on thin brashy wood, or other things which will rot and go together, cannot assist in the drainage. The stubble, hay, or grass, which is placed upon the tile will remain, although wet, for years, and the only real use of it is, that in the first instance it allows of the soil settling down naturally upon the drain, without allowing any of it to pass through, and yet at the same time suffering the water to pass through it; and it also prevents the breakage of the tile in returning in the soil by any stones which may be in it.

It is as important to get off the water rapidly, as it is to drain at all; and by the adoption of the above simple plan, this will be accomplished. All the water which falls in twenty-four hours upon the strongest land, will be completely and entirely removed in about two days after rain, where this mode of Draining is adopted.

Having got, by the foregoing process, a depth of twenty inches into strong clay, the tile will occupy four inches, and not more; the size of the tile for this description of Draining being three inches by four inches in the opening, which will be as large as is requisite, and not too big for all the parrallel drains. The tile will then be fifteen inches deep; and supposing it ever should be the wish of the occupier to subsoil this land with the depth of soil I have before stated; if he goes twelve inches deep—and this will be more than needful—he will then be three inches above the tile, (if he scours the furrow at the same depth,) and this without allowing anything for ploughing down and laying the land flatter, which will give a few inches more; as it frequently happens that land laid ready for Draining, is ploughed up to the ridge twice before the work is done, and in many instances where the clay is very stiff and retentive, nearly all the soil is ploughed out of the furrow towards the ridge; and this is more clearly shown or seen when the field is in corn. The furrow for a yard on each side will exhibit nothing but thin spindling straw, from a foot to two feet high, and the ears on the top of it in like proportion, whilst the higher part of the land will exhibit as beautiful a specimen of fine straw and corn as can be grown.

Having, in the foregoing remarks, stated twenty inches to be a sufficient depth, I do not mean this to apply to all descriptions of land, as in one of the fields shown at the end, the drains there vary from eighteen inches to two feet six inches, the whole of the underneath subsoil or strata consisting of a mixture of shells of all descriptions, in a tolerably soft state, for several inches, and below that, at depths varying from one foot to two feet, a strong incrustation of shells, so hard as to require the use of picks 12lbs. in weight, the points of which, when well laid, would scarcely last a day. This again varied in thickness, and below the shell stone was a thin hard brashy stone in layers, with open measures, admitting the water from underneath at various depths. Owing to the rise and fall of the lands at the greatest depth in most parts of the field, the soil or subsoil stirred would not exceed eighteen inches, whilst in other places there would not be more than eight inches, on account of the shell stones, and therefore, with the exception of those drains made deeper for taking the bottom water, it would be perfectly unnecessary, and throwing money away, to cut them deeper than I have stated.

When I first saw the field above alluded to, it was after heavy rain, and the furrow stood six and eight inches deep in water; and upon that part where the outfall is cut, there were places in it up to the knees in water. I have walked over it since it was drained, and the land, although heavy rains have fallen upon it, has been as dry as a bone. The quantity of water discharged at one of the outfalls is a tile five inches and a half by five inches, within, quite full at times, (after rain,) and at other times a running stream of one inch deep to two inches in the tile. The other outfall, (the dip of the field

H

requiring two,) discharges a tile of the same size, half full. The result of the cropping upon this piece I am not able to state, it being the first crop since it was done, and I have alluded to it in order to show the depth necessary to drain it. The cost of draining this piece of land was about £49., or an average of £5. 10s. per acre.

If the soil of the field is a marl or loam, and the subsoil open and porous to the depth of thirty inches, and of that quality that it may either be stirred up to the top or incorporated with the top soil; it may then be advisable to put the drains in at thirty inches. But it is then a question as to whether, when you deepen the drains in this description of land, you may not at the same time extend the width of your lands; and as it is really important to save in some way, the lands should be set out at ten yards, or from that to fifteen yards apart, and no fear need arise as to the water drawing off if the drains are but laid with a proper fall, so as to carry off the water freely—this will happen generally from the natural inclination of the land; but if it is too flat to draw off naturally, the fall must be got in the drains, that is, from making the ditch receive your water as low as you possibly can, and also regulating your outfall drain in the same proportion. And here it will happen that it is important to see the drains run before the tiles are laid in; and if not sufficient to see of themselves with rain, water should be carried to them. The experienced Drainer, who cuts the drains, will be able to tell pretty near, but water is the safest and surest method of trying, before all others, as it is impossible to err when you see them run. Of course, according to the number of chains in length, which your drain will run, there should be an allowance in the distance, of one or more inches, for the fall, in a chain; therefore at the outfall, supposing your drain to be twelve chains long, it will be three feet deep, and the top end of your drain two feet, to give an average of thirty inches; but if you wish the drain to be that depth at the top end, you must, as a matter of course, deepen your outfall, (if it will admit of it,) so as to get the extra six inches there. In the above I am speaking of very flat or perfectly flat land, as with land having a proper inclination from the bottom to the top, the above will not apply; and in allowing one inch in a chain it may be considered and found to be very trifling, but the water will run off with this fall; and I mention it, as it is not only a point very much lost sight of by some, but by numbers never even considered at all.

Upon all the low levels of Lincolnshire, Cambridgeshire, and those flats lying next the sea, along the eastern coast, as well as in other parts of the kingdom, the above description of

Draining will apply, where the land requires Shallow Draining; but if the soil in many of these districts (and there are many where it is,) is so open and porous as not only to receive the water from the top, but also allows the water to soak through from the side drains or dykes; and where the great outfall waters are carried upon a level for miles, only one range or depth can be adopted, and the parties who drain in these districts, should adhere to the rule laid down of having but one outfall, and they must be guided by the depth of their side drains as to the lowest depth which they can place their tile in the ground, to be of any benefit for draining the land from surface water. In many parts a foot or eighteen inches can only be obtained, whilst in others two feet and even two feet six inches may be got, for the purpose of laying in the tiles; and in the first instance, subsoiling this land would be out of the question, whilst, in the latter, it falls, even then, six inches short of the rule laid down by many Drainers, that these drains should all be three feet drains: and here the absurdity of such statements is very apparent.

The same rule applies here too as in strong clay, when the bottom is soft, as in all cases where this happens, or that the land is open and porous, it will be advisable to lay flat tiles; but when the tile comes on the top of the strong clay, this is not necessary. One point, too, connected with the outfall drain, is important, and that is, always to have the tile sufficiently large; the tile No. 3, which is five by six in the opening, should be used for this purpose, not that half an inch variation in the size of the tile moulds will make any great difference, but it is as well to have them large enough, and no great additional expense is incurred in the execution of the work. Back water too up the tiles for part of the distance cannot materially injure the drains, so long as it keeps to the range of the tile, but if it gets above the top of the tile, of course all the land will be alike injured.

I have seen instances of drains open that have been described as two feet, when, if you had examined them from one end to the other, they would not have averaged more than 18 inches. When you are employing men by the day, (a very bad system at this work), it does not signify, because you can alter them if you choose, supposing you measure them and find them shallower than you intended; but as it is always best to employ them by measure, it is important to examine them, in order to see that the work is performed fully, and that you may not be paying them more than they are entitled to.

My reasons for having not only this, but all work connected with Draining, done by measure, instead of by the day, is, that some men will do half, and, at times, nearly as much more as others do, and if one or two idle men are employed with others, they are apt to infuse some of the same lazy spirit into their comrades; besides, when a man knows what he is working for, he can regulate his time accordingly, and I have had men who, whilst one was earning 4$s.$ per day, another was only earning 2$s.$ 6$d.$ How exceedingly unfair and unjust it would be to put these two at the same price: there is just the same difference too in the value of this work, when seen properly, as there is amongst the different members of any particular trade; whilst one is exceedingly clever, not only in digging but in casting, another will take more time over it, and not do it near so well. Thus I would endeavour to lay down the principle of paying the labourer according to the work he performs. It has been objected to by some, that this class of men make a great deal of wages, and I would answer to this, that their work is exceedingly hard, and having their own tools to find, it is necessary that they should make more than the ordinary class of labourers; and there cannot be a doubt, that if they make double the wages, they do their work in half the time that the ordinary labourer can, and are therefore worth double the amount.

There is one thing connected with Shallow Draining, as now performed, which I cannot pass over without comment, and that is, the practice of ploughing out the furrow for a certain depth instead of digging it. My plan, as before stated, is to have the drains set out with a *line*, and I should recommend this mode, not only as being more perfect, but far superior to any ploughing which either has been, or can at present be, adopted. There are not many fields where the lands are so set out, as that the lines of drains can be ploughed straight through, and it is very often necessary to take cross drains, (besides the outfall,) so as to secure the water all running one way. And what is gained by this horse labour?— the soil or clay can neither be so well got out or so easily removed from the clay, as when dug and cast; and taken altogether, it is a slovenly way of attempting to do that which ought to be done by manual labour, which is performed at a trifling amount of additional expense, but which is far better executed than it can be by any ploughing in the present day, when so many schemes are afloat for destroying hand labour. I do hope the owners and occupiers of the soil will not encourage this lazy and inefficient mode of doing the work, in opposition to the industrious labouring class.

In addition to the above remarks, there is one thing not a little extraordinary, and that is, the strong degree of prejudice which exists in the minds of workmen, as to the way and the depth that drains should be cut. I do not mean the regular Drainer, (though even some of these are stupid enough in their ideas), but that class of workmen who have done some little under the directions of farmers, and have not followed it altogether. I have known instances of parties who have professed to know how Draining should be performed, completely led away by the ridiculous notions of their men. I knew an instance of a Gentleman in Nottinghamshire, who was a great advocate for Draining, and professed to know how it should be performed, who set out two pieces of strong land to be drained, and actually allowed the men, whom he had employed to do the work (and who had been draining, or rather digging, a few months elsewhere,) not only to alter his plan, but did it contrary to the rules he had himself laid down, (of having but one outfall), and let every drain empty itself into the open dyke, thereby making a dozen or two of mouths where two only were required: and this was not all; instead of putting in the tiles at the depth he had stated, *viz.* thirty inches, I examined these drains at the time, and found that not one of them exceeded twenty, and this was intended to be subsoiled.

This is not a solitary instance, for I saw some tile draining properly set out, and the tiles put in in a workmanlike manner, afterwards filled up, at alternate lengths of a chain together, within four inches of the level of the surface soil, with stones, under the direction of a Bailiff, in the county of Northampton: but this was not at all surprising, as the same party afterwards actually put tile drains under a road at given distances, (in the same manner that some would drain a field,) on purpose to dry the road, the same road having a ditch on each side of it, and a fall from three to five feet in a chain.

These are lamentable instances of parties performing, or endeavouring to perform, what they really do not understand, and are to be deplored, as not only causing expenses which are needless, but actually throwing the property of others away. The man who really intends to have his land drained, should lay down his own plan, satisfied that he is right, unless he gets the assistance of some one who has had practical knowledge of the work: and as no very great difficulties present themselves in the execution of this mode of draining, there are few men who cannot make themselves masters of the work, and who, by attention to it during its progress, may see it efficiently performed, and the land completely drained And I would particularly recommend the presence of the master when the drains are to be tried, and the tiles put into the ground, as it is of no use going to look at them when covered over: they may run part of the way and not the whole of it, and it would be a very clever man indeed who could vouch for land being "thoroughly" drained, by merely looking at the mouths of the drains before closing up the outfall; and it is here, therefore, that attention is particularly required, to save both the labour and tiles from being thrown away, as well as a portion of the future crops to be grown upon the land.

DEEP DRAINING.

SHOULD BE EXECUTED BY THE LANDLORD.—MODE OF DETECTING DEEP WATER.—TRIAL HOLES TO BE MADE BEFORE COMMENCING THE WORK.—REASONS WHY PRACTICAL WORKING MEN SHOULD BE EMPLOYED.—THE LEVEL OF THE WATER TO BE TAKEN AT STARTING, AND CONTINUED TO THE END—TILES, HOW SECURED.—PLANKING AND STRETCHING THE DRAIN, TO SECURE THE SIDES FROM FALLING.—WHAT TO USE FOR THIS PURPOSE.—WHERE TO FINISH THE DRAIN.—BRANCHES INTO THE MAIN DRAIN OCCASIONALLY REQUIRED.—SIZE OF TILES TO BE USED.—WHERE FLAT TILES CANNOT BE PROCURED, BLUE SLATE MAY BE USED.—CARRYING OUTFALL OVER DRY GROUND.—DRAIN EMPTYING UNDER WATER INTO A DYKE OR RIVER.—REFERENCE TO ONE EXECUTED.—DEEP DRAINS EXECUTED, WITH REMARKS UPON, IN NOTTINGHAMSHIRE, LINCOLNSHIRE, DERBYSHIRE AND NORTHAMPTONSHIRE.—NECESSITY FOR USING THE BORING RODS IN DEEP DRAINS.—VARIATIONS OF CUTTINGS, AND QUANTITIES OF WATER DISCHARGED FROM DRAINS.

BEFORE I proceed with the Remarks upon this important subject, I must premise them by stating, that this is to all intents and purposes, the *work of a Landlord and not of a Tenant*. The great expense and the magnitude of the undertaking, in nearly all cases where it is required, are far above the reach of the tenant, however much he may disposed to enter upon it, and the only way in which he could be secured, would be by granting him a lease of his farm, subject to conditions to be entered into at the time, as to the probable expense to be incurred, or to the term or duration of the lease, as to the time when the tenant would be supposed to have derived the benefit sought for, by the investment of his Capital in the land; or else, if he is tenant from year to year, an allowance or proportion of the expense of executing it might be made in the rent every half-year, until the whole is paid off, so as soon to get rid of the incumbrance. But I think it would be much better for the landlord to do it, and place, as before stated, a per centage upon the rent, as not only are many of the tenants deficient in the means, but it would be taking out of their hands for a time, that which ought to be employed in the management of the farm; and even with a lease, it would have to be considered in the rent, (as well as the term to be granted,) as if an outlay of £200. or £300. is to be incurred, the yearly interest of such money should be taken into consideration as well as the principal, and also the probable length of time required for executing the work. In addition to these reasons, there is another and a very powerful one, why it should be executed by the landlord, and that is, that up to the present time, not only is it very little known, but there is no Valuer who could venture to put a proper value upon it, either as to first cost, or the term to be allowed for wearing or running out; such being the case, and being a point to which the attention of Valuers has not yet been drawn, many having never seen it performed, or know indeed anything about it, it would be a matter of impossibility to do that which would secure both the landlord and tenant, the one from making too great an allowance, the other from having too much to pay. Under these circumstances, I should seriously recommend the adoption of it by the landlord himself, and not only for the reasons before assigned, but that if done at his expense, or by parties paid by him, it will, in all probability, be more likely to be done properly than by the tenant; for, although he may be dispoeed to do it *well* if left to him, I have known many instances where the expense of "Shallow Draining," has deterred parties from going on, even when allowed the materials necessary by the landlord, and where, had they executed the work, the second or third crop would have paid them every farthing. With these facts staring us in the face, I think it must, at present, certainly be considered a *landlord's job.*

Of all descriptions and modes of Draining, "Deep Draining" is the most difficult, and least understood, and it is, at the same time, of more importance than any other kind of draining which can be done. It requires the party undertaking to perform it, to be perfectly acquainted with the rise of water upon the land, and to be able at the same time to carry it off. It is of little use to try and drain land by Shallow Drains, if the water ascends from below the level which those drains would take; and it is also a matter of impossibility to carry off water by Deep Draining, where the strong and retentive nature of the subsoil refuses to allow it to pass into the earth, but keeps it upon the surface like water in a vessel: so that it not unfrequently happens, that land requires both descriptions of draining before it is complete. And, first, the question arises how this is to be ascertained, where land is subject to wet from bottom or spring water, and not shewn by its constant running, but by the wet and boggy state of the ground. The best time to discover the outbreak of it is (if in tillage) when fallow, after continued dry weather, when the line of wet will be as plain and distinct upon the land as the ditch running by the side of it, or the furrow ploughed across it. But if it is grass land, the outbreak of the water will be indicated by the appearance of rushes upon the land, not over strong in some cases, but in others very luxuriant, depending in a great measure upon the nature of the soil, as rushes, like other plants, generally do best where the soil is deep. There is also another mode of detecting deep water, and this will be found even upon high land, and that is a strong mixture of small shells in the soil, consisting sometimes of a black mould, and at other times of a mixture of black sandy gravel, sometimes rather flinty, intermixed with the shells; the appearance of these are a sure indication of deep water, wherever they are found. Another, is the strong luxuriant appearance of what in Nottinghamshire, is known as toad pipe, growing out of the ground to the height of a foot or eighteen inches, in the form of a feather, only that its side feathers or leaves extend all round it. This plant rarely grows but where it is supported by water at a great depth, and may be considered the Drainer's friend. After discovering any of these strong indications, to ascertain what depth the water will rise, and how low it lays, it will be necessary to dig three or four holes above the line of water or rushes shewn, not upon the wet land, but upon the dry: let these holes be dug sufficiently deep, not to give over when the party digging thinks he is deep enough, unless the water forces him out of the hole, but let him go from five to eight feet, and it is quite possible they may get into a hole where the water will rise at five feet, and that at eight feet there may be none, and yet within a few yards it may break out again.

Supposing you have found out where to drive your drain, it is not of the slightest use your proceeding with it, unless you have got a man fully qualified to put in the tiles, as the *onus* of the work, in a great measure, depends upon this. I speak of tiles here, as I do not know that any sensible man would think of Deep Draining with stone: you might as well dig your drain out, and throw your stone in, as you would fill up a hole, unless you were going to make a culvert, and even then, in the ordinary course of Deep Draining, the impediments are very great; as, if the supply of water is large, and in a running sand, which frequently happens, the difficulty is to get through it, and can only be effectually accomplished with tiles. Any man may dig out a drain down to the bottom, but not one in a hundred of those who do the digging is capable of putting in the tiles: this is as complete a science as that of a slater slating the roof of a house, or a mason laying the bricks and stone to build it with. It is the foundation of the work, and every thing depends upon its being done well. No bustling or hurrying in laying the tiles will secure this.

The man who lays the tile requires considerable experience before he should be allowed to proceed to do this, and should work as second man at the bottom for some time, to make him acquainted not only with the danger he has to encounter, but also to see the tiles laid, and occasionally assist in laying them. It was the custom of Tebbet, not to employ his men in laying tiles, until they had been a year or two at the cuttings, and then those whom he selected were men with coolness, skill, and judgment. The party laying the tiles requires to work in the running measures with quickness and dispatch, his eye invariably looking to the walls which encompass him, so as to watch whether a portion will enclose him, or whether he will be able to get one or two tiles in, before any portion gives way, which is frequently the case; he must also have his bottom or level perfectly true, and pull out the plug from his last tile, and try the water, before he lays his flat tiles. In working his drain forward, he must pass all the earth over his shoulders which it is necessary to remove, upon that portion which is complete, and at the same time keep his stretchers in where necessary, so as to prevent its falling in before the tile drain reaches it. Another important consideration, too, is, that in the course of his workings, he will be very likely to lose the water, and it may be requisite to use the boring rods, and try for a lower level of water; and as you cannot always be with him, he should have some knowledge where these borings are to take place, as, without boring at times, the whole of the supply of water may not be thoroughly secured; therefore, if your man is not acquainted with this, his work will still be imperfect. It does not follow that the man who executes the work knows anything of the way in which it should be set out, but still, from seeing it done, he is better able to judge than those who have never seen any set out at all. If he has a quantity of work to do in any one parish, he may make himself, in a very short time, master of the way and the measures at which the water rises, but the strata of the earth varies so much in other parishes, that the working of no two drains will be alike, varying in the cutting every ten or twelve yards, (and the most experienced Geologist would be puzzled to give a correct section, unless he actually saw the workings as they proceeded). Here then he would fail in his object; and these are the places where the observations, as to the appearance of deep water, before made, require attention. Another essential point is the carrying the tiles over places where the bottom is soft and porous; particular care must be had here, that you have it made so secure, that it shall not purge up into the tile, or blow up the silt, so as to cause it to fill up. If your man has not had experience in laying tiles, here too he will shew it; and even the practical man is sometimes so obstructed by these impediments, as to be nearly overcome. I have had and known instances where the bottom would not stand, and the sides have given way, that the whole party have run from the job, and great difficulties have been experienced in being able to carry it forward at all, and in these cases great courage and determination must belong to your bottom man, or you will fail in your object.

Having discovered the source of the water, and the depth which it will be requisite to go to find it, the next point is to select your outfall, of which, in some places, you may have a great fall, and in other places very little. I shall proceed, supposing you to have a good outfall at a level of a foot above the bottom of the drain, or immediately at the top of the drain water, (that is, the open ditch for the level drainage of the parish,) where your water is to empty itself. You must then proceed, with your drain open, until such times as you meet the water; the object of which will be very apparent, because you can then see whether your tiles are put in properly at starting, and without this, and a careful attention throughout, no man ought to begin this work at all, as an inch too high, or half an inch too low, will spoil the undertaking, though much sooner from being too low than too high. Supposing you pass the level over it to try the tiles, what an expense is incurred if you have ten or twelve yards to take up again, because it is improperly put in, and it is no unusual thing for this to require doing, without even losing the level, from a quantity of silt escaping into the drain, at the time of laying or putting in the last tile.

The Drainer having proceeded in cutting to the point where the water makes its appearance, and having got sufficient water to proceed with, the next step is to place in the mouth of the tile a piece of clay to stop the water, an inch in height, so as to drive the water back to that level; this will not prevent the water discharging itself at the mouth of the drain, as the water keeps rising in the drain it will flow over the clay. The extent necessary to keep the drain open will sometimes be three or four chains, depending in a great measure upon finding water. If the bottom is strong, it will not be necessary to put in anything but the flat tile, but if it is porous it will be absolutely necessary to use refuse hay or stubble, and the finer the hay, or rough grass mown for the purpose, the better; and this should be continued throughout the work. Having put in the tiles, as the work proceeds, each tile should be well lined on the sides, and carefully covered on the top with sods, which can generally be got upon the land to be drained, and if not, must be carted to the place for the purpose, as the tile must be made secure, and at the same time perfectly covered, so as to prevent anything from getting either into or upon it to break it. The sod acts as the filterer for any water which may come upon the top, and the drain itself takes that which comes from underneath it through the joinings or interstices of the tile. Upon the top of the sods may be thrown a few shovels full of earth, from the surface, and then he may proceed, after first putting in a plug or hay wisp, so as to fill up his tile, and make it secure; and when the drain gets deeper, to place a sod before the tile, in addition. The object of this plug or wisp is, to prevent the water and the silt from passing into the drain during the time of digging, and also the water from backing after him, until he is ready for the next tile. He then proceeds to clear the ground forward, for the purpose of laying more tiles, and casts his stuff behind him; his guide being now the level of the water which he has in the drain, and this he keeps, unless any sudden change occurs, unto the end. It is possible he may cross a place without any water in it, and that he may be deficient in his supply; when such is the case he can generally get a few buckets full to help him forward, from the holes dug to guide him, until he meets with the water again. It will sometimes, towards the close of the drain, be found advisable to lose the level, and raise the drain higher, but this is a dangerous experiment, and one that ought not to be tried but by the practical Drainer, as, should the water be strong, and any silt or dirt get into the tile, it will be apt to lodge when it meets the level of the drain, there not being then sufficient current to carry it forward. Every time a fresh tile, or tiles are laid, the last tile should be cleaned back with the hand, so as to prevent anything from lodging in it, or else a stick about a yard long, with a bit of wood upon it, to act as a scraper, passed into the tile to ascertain that it is perfectly clear; for, in spite of all the care or pains which can be used, some will pass with the water when the tile is opened.

The work will then proceed, until the drain reaches the depth of seven, or, in some instances, (if the sides are solid,) to eight or more feet; but if not, it will be necessary to plank and stretch it, so as to prevent the sides falling in; and this is another instance where the skill of the workman is required, as from habit he will know where to place them, and not to have too many in to crowd or obstruct him in his work. It may here be remarked, that the old maxim will apply, of "a stitch in time saves nine;" as, if one or two are put in beforehand, a great deal of labour may be saved by preventing the sides falling in and burying a portion of the work already executed.

Still there are (I have known many) instances where, notwithstanding all the foresight and caution used by the workman, the ground has given way for yards together, and tons of earth have had to be removed off the place, where a few minutes before it has been ready for putting in the tiles.

The planks may be slabs of timber varying from five to ten feet in length, and about nine or ten inches broad; but if any great quantity of work is about to be performed, deals cut into lengths are the best, and as cheap as anything which can be used, and they lay to the sides much flatter than slabs will. The stretchers are best cut out of oak poles, about three or four inches in diameter; they may be cut from other poles, such as larch or spruce, &c., and will answer the purpose, but are not to be depended upon like the oak, (unless it is larch cut with the sap in); these should be cut into various lengths according to the cutting of the drain, in width from two feet six inches to five feet.

From this time, that is, from eight to twelve feet, and from that to twenty feet, considerable skill is required in the second man, in disposing of the materials in his road. The man laying the tiles is now working below at a greater depth, and it is impossible for much more than half the soil now cast out, or to be removed, to reach the top of the land; consequently the second man must cast the whole of what he digs over the head of the bottom man, and do it so free and clear as not to cover him, and also throw it clear through the stretchers as well, otherwise it would fall back and incommode him in the progress of the drain at the bottom.

To those who have never seen "*Deep Draining,*" and talk of it as a mere matter of digging, (and there are hundreds who have not,) now is the time to see the work, as one both of strong manual labour, and skill too, on the part of the workmen; and none but men well accustomed to it, would either be able to execute it, or have the courage and resolution to face it; and it is here too, when the work has reached the above depths, that the skill of the man is required in properly placing his tiles. It very often happens that he can lay but one tile at a time, and rarely, if ever, more than three, owing to the water which he has to meet, or the dangerous state of the sides of the drain, as he proceeds, which fall in upon his work.

It does not follow that all deep drains should be carried to the depths I have named, but it will happen that the termination of the cutting is fourteen or fifteen feet, or perhaps more. A great deal depends upon the line of ground you have to pass over, and where the outbreak of the water commences; as, if it continues rising from the outfall to the head, it will require the drain to be very deep; and, on the contrary, where the outfall extends only a chain or two up the land, and then turns across a level piece of land, or with slight elevations, the average depth will not be more than nine or ten feet from the turn, or even less.

To know where to terminate, will be generally at the end of the outbreak of the water in the line adopted, or, if this should not be considered sufficient, if a deep hole is made, a distance beyond where the line is to be carried, and it is found that as the drain proceeds this water is gradually drawn off long before reaching it; this will sufficiently indicate to any one that the object of the drain is accomplished. It rarely happens that more than one line is requisite to drain a district of land, if properly set out, and can be pursued; but there are times when you are compelled to stop short, and leave part of the evil uncured, from the land adjoining not being your own, and your neighbour with no power to assist you in carrying it forward; or else obstinately persuaded that yours is a foolish scheme, and will not answer, and therefore he will not afford you any assistance in promoting your object. These are obstacles to improvement, and are to be regretted when you meet with them.

But to return to the line of Drainage. The inequality of the land will present a valley to the right or left of your drainage, as the case may be, where the outbreak lays further up, and in cases of this sort it will be necessary to carry a straight cut from the main line into the very heart of it;

to tap, and send it down to the same outfall; and when this is done, great care is required at the junction of the tiles, to admit of both waters passing freely into each other; and it is advisable, before proceeding straight, to give a slight curve to the branch drain at starting, as in cases of this sort, the current of the straight drain will then carry any slight deposit of silt more freely. To ensure the full completion and perfection of the work, due regard should be had to the size of the tiles requisite for its execution. A great evil has often arisen, and I am sorry to say I have fallen into it myself, not from inclination, but from inability in procuring the tiles of the size required, and therefore obliged to use smaller ones; and wherever you have a large body of water to contend with, your tile should be of the Nos. 5, 6, or 7 size, at the beginning. It is no use putting in a large tile when your outfall tile is full, therefore begin with the large, and when you have passed the principal supplies, then you can use the smaller numbers. No deep draining ought to be begun with a less tile than No. 4, which is six inches by seven inches within, and the flats for these should be made to fit them. If flat tiles cannot be procured, I would recommend the use of the strong blue slate, which can be obtained in most districts, and will be found as useful, though not so economical, as flat tiles. I have lately used them in a drain at the rate of £6 per thousand, of twenty inches by ten inches. There is no breakage scarcely in them; they are quicker laid, as a man may lay two slates whilst he is laying four tiles, and then not get them to fit so well. Besides, there is economy in the time saved; the four tiles requiring more time, the earth may come in upon them whilst in the act of laying them, and bury them, so as to require part of a day to get at them again, and thus retard the work,—the additional expense will probably be saved in the time gained. I do not recommend them for general use; but there is a decided objection by the brickmakers I have had to deal with in some places, to make a good-sized flat tile: one, and a very valid one too, is, that the clay will not hold together, or else that they break in the burning; it is, therefore, in cases of this sort, and to get over the difficulty, I would recommend the slate, so as not to throw any obstacle in the way, to prevent carrying the work forward.

In Northamptonshire I have used, and seen used, the Easton or white slate; but they are very bad for the purpose, being very uneven on the surface (unless the very best), and require a good deal of labour to make them fit close together, and it is highly requisite that they should not only join, but also lay solid, so that nothing may disturb them when once placed in the ground.

It may be necessary to carry the outfall of a drain over ground where no outfall for water ever existed; this of course is an awkward task, and one where care is required. Not having water at starting, it will be requisite to have some to try it as you go on, and see whether you are proceeding right or not, as the soil may be dry and porous, and instead of your water passing out at the drain's mouth, may make its escape into the earth, and appear in another place, where you do not want it.

I cut a deep drain some time ago, and had this difficulty to contend with: my man, a thoroughly practical one, was sure he could cut through the land and lay the tiles without any water. I, knowing his ability, consented to it, and he had completed about four chains, and got to a fine run of water, at least an inch deep in a tile five and a half inches within, but at the mouth not one drop, the tile being as dry as when first put in. I directly opened down to where the tile began to run a little, and after tracing it, I found the whole disappeared into the earth. I again opened the outfall all the way until we met the water, and put in the tiles (both flats and arches) with clay, the distance of two chains, and then the whole appeared at the mouth, and has continued so ever since. The drain discharges about twenty gallons per minute. This is one instance of breaking through dry land for an outfall; I have seen others, but they do not occur often.

There is another mode of getting an outfall, should the circumstances of the case require it, or the level of the water in the ditch into which you wish to carry your drain, stands higher than you want it, say a foot or more, and you have no means of lowering it to the depth required; but this is a plan or system I do not recommend, and it seldom occurs. Dam off the dyke water, and do not let it interfere with your drain at all, but proceed as if the whole were perfectly clear, and continue to the end: of course, if your water should rise from the drain into the dam, it will be necessary to keep relieving it by throwing over into the dyke, or else you will have too much back water up the tile drain. When your drain is completed, you may remove the obstruction, and let in the dyke or ditch water, and you will find that if your current of water is strong, that it will pass into the stream, although under water, in the same manner that it would if the drain's mouth had been above the level of the water, (though not with the same visible force). The back water, which you may suppose will prove injurious, will certainly do nothing of the kind; the force of water in your drain will drive it forward, and shew the run (when the water is perfectly clear) in the ditch or river into which you take it. I know but of one instance where this description of outfall occurs, and it is in Thoresby Park, Nottinghamshire, in a drain cut by Tebbet;—it is into the bed of the river, and I have many times viewed it, and seen the water pass along into the current of the stream, and flow, to all appearance, as rapidly as if no water were passing over it. The main object of the outfall of this drain was to prevent an immense length of cutting, to get a lower level, but I have no doubt the master Drainer was satisfied it would answer the end intended, or it would not have been executed, and in this it was evident he was right, as years have now elapsed since its completion, and it still continues to perform the object desired.

A large quantity of Deep Draining has come under my notice in the counties of Nottingham, Lincoln, Northampton, &c., and I shall here give an account of a few of the drains which I have seen cut, and others executed under my own superintendance, together with the causes, modes of doing them, and other observations in the progress of the works, which will, I trust, be not only interesting, but beneficial, in stimulating others to pursue this plan of Draining, where required.

Amongst the first drains to which my attention was directed, and which made an early impression upon me, were those in execution under Tebbet, at Clipstone, Welbeck, Carburton, and Edwinstowe. The first time I saw a deep drain cut, after carefully examining the way it was set out, and the progress of it, I felt satisfied that it was idle to talk of draining land by "Shallow Draining" only; years have tended to confirm that impression, and also satisfied me, that it requires experience to make a man a "thorough Drainer," and that every man cannot take upon himself to say he can thoroughly remove the cause of deep water, unless he has given great attention to it. I could, at this time, enumerate several large and extensive drains, cut under the direction of very clever men, (as regards other pursuits) which were miserable failures; for the sake of the parties and their employers, I shall not name them. The work has since, in many instances, been done over again, and answered the purpose.

In addition to the drains in the parishes before stated, there are some cut at Palethorpe, Eakring, Kneesal, &c.; and amongst those at Eakring, is one which supplies a farmstead with water from the drainage of the land. The farm in question is one of those outside farms (many more of which are required all over the country) built for the better occupation of the land; and was built, as is too often the practice, without considering that it would be necessary to have water to it. The farm itself was naturally wet, and the artificial means employed to remove the water from the land, were also the cause of supplying the deficiency to the buildings; and it not unfrequently happens that this deep water may be made available upon farmsteads, for more purposes than that of watering the buildings, as very often it may in its transit, be made also useful in giving a supply of water to the stock upon the land, by means of troughs placed to the level of the water in the drain, and conveyed either by outlets from the drain itself, or else by being put at the mouth of the drain, and the overflow allowed to escape into the ditch adjoining.

About ten years ago a deep drain was cut in the parish of Kneesal, Nottinghamshire, in a farm occupied by Mr. Townrowe: it was cut on the hill side of a field, from six to seven chains long, at an average of about six feet, by Samuel Smith. A very considerable supply of water was obtained, which effectually dried the land, but there was nothing in it out of the ordinary mode of cutting, until finished, except the removal of the water in a very singular manner, and to which I shall hereafter have occasion to refer at length, under the head of "Boring."

In the year 1835, in passing over a farm in the lordship of Hemingby, in Lincolnshire, in company with William Simpson, Esq , the then Agent to Earl Manvers, we were astonished at the appearance and indication of water in a field in the occupation of Mr. Todd, (a very worthy

"Old English Farmer" residing there, and whose family had lived in the place for generations). The land in some places was tolerably dry, whilst in others it was a complete jumping bog, particularly on the hill sides, and at the top of the hill. In many places you could get upon a piece, and by jumping upon it, the ground for eight or ten yards round would shake and tremble, as if the whole were about to give way, or sink in. On inquiry of Mr. Todd if any steps had been taken towards draining it, he said, "he had been trying for years," and that "he had put in hundreds of tiles, without being able to cure it, and he should be very glad to have it done, if he paid for the doing of it himself." Samuel Smith was sent to effect the drainage, after a description of the land, &c., had been given to him. An open drain was run up the valley into the field, and into this was brought three outfalls from the hill side. This was necessary from the formation of the field, and after the three drains were cut, and run into the main drain or ditch, the whole formed a supply sufficient to turn a water mill, could one have been placed in the field. In examining this field two years afterwards, the land had become so solid, that it would bear a loaded cart or waggon over those places, where before it would scarce bear a child; the rushes from being strong and luxuriant had dwindled down, and were gradually disappearing, and a new herbage was springing up on the part once a bog. The expense of Draining this field of rather more than twenty acres, was about seventy-five pounds.

Similar results to the above also occurred in draining the grass field, adjoining to Stainsby Hall, on the other side of Horncastle, in the occupation of Mr. Charles Lichford. The depth of the cutting in the deepest part was about twelve feet, but in drying the land, the pond in the field, and also the well adjoining the premises, were dried; and it was necessary to deepen it, to get the premises supplied. The cuttings of this drain were partly through a sand rock, and part running sand. This field, together with other parts of this lordship, are now perfectly dry, and places where it was almost impossible to get upon, are now perfectly free from water, and producing a most luxuriant and beautiful herbage.

In the year 1836, a piece of land in the hamlet of Sothal, in the parish of Beighton, (before mentioned,) within five miles of Sheffield, was drained. The land injured comprised about two acres, and was a complete bog, unfit for anything, and without the possibility of getting anything upon it. The description of Draining was quite new in this part of the country; tiles having been but rarely used, the great quantity of stone found in all this part of the country being used for Shallow Draining, and Deep Draining, from the nature of the ground, or hilly districts, being rarely attempted: and in addition to this, large bodies of spring water are cut off from the outbreaks, by the drainage necessary for the coal mines, and other minerals in the land.

This bog lay upon the low side of a hill, (not the lowest,) being several feet above the level of the road below it, and immediately above the line of cutting for the drain was a rock of stone, extending in a line directly across the hill. No stone scarcely was found in cutting the drain, the whole line being through sand. The land is now perfectly dry and convertible, whereas before, it was to the occupier an intolerable nuisance. The cost of executing this drain was about £35, and was to the tenant like giving him a piece of land, as it lay within about three hundred yards of his farmstead, in the centre of land worth about £2 an acre to rent. The result of the drainage of this piece, and adding it to his farm as good arable land, was far better than any return would have been to him in the shape of a per centage upon his rent.

The next drain I shall mention is one cut across some fields in the parish of Kneesal, near the Greenwood, and conveyed on the level across several fields, successively watering them in its progress to the farmstead of Brocoli, in the parish of Laxton, where it runs into a well made for the purpose of receiving it, and from thence the overflow is carried off into one of the open drains adjoining. This drain was executed by Samuel Smith. It was absolutely necessary to dry the land through which the drain was cut, and before this drain was executed, this farmstead had no water but what was carried up hill, (the premises standing upon a high hill,) from a well in the valley, at a considerable distance from the house, except water which was caught in a ditch above the farmstead, and supplied the yard, until it was exhausted by the stock and drought together. By the aid of this drain, the farm was well supplied with water, and the land thoroughly drained through which it passed.

I commenced, in the year 1841, a deep drain upon some strong land in the parish of Swinderby, in the county of Lincoln, and running very near the boundary of the counties of Lincoln and Nottingham. The soil of this land, principally a strong retentive clay, upon a subsoil varying from strong blue clay to beds of gravel, intermixed, and extending across the range of the fields, as described in the plan annexed, No. 3. A portion of this land was shallow drained, and was certainly done in the most slovenly manner I ever saw draining done. As I had occasion to take some of it up, partly from necessity, and partly from curiosity, upon that part which was drained it was quite evident very little good had been effected, and it was apparent that other means would be necessary, to effectually dry the land. In the fourth field, considerably elevated above the rest of the land, was a pond, never known to be dry by the oldest man living in the parish, and, from vulgar notions held by some living in the parish, it was stated, that it could not be dried, as it was one of two or three in the county which were obliged to be kept open for the use of the public, or the county at large, and to which they were entitled at all times to have access. I was sure that this pond was the cause, not only of immense injury upon this land, but also upon the land immediately adjoining. I therefore dug a trial hole, about one hundred yards from the pond, as near the boundary of the land as it would be prudent to go; the hole was about seven feet deep, and the water immediately began to rise in the hole, and in half an hour from the time of cutting it, it filled the hole to within twenty inches of the top, and there continued, not one drop having fallen into it, clearly showing that there was a heavy pressure of water from the hill, and which must make its escape into the land at other places.

I commenced the drain in the third field, and set out the line straight for the pond, intending to cover as much land as possible, and in case of more water than that indicated at top, breaking in from the *west*, in the line of drainage, I should be able to secure it all. The cutting was through blocks or measures of clay and gravel, across two fields, until it joined the second, when it entered a thick bed of blue clay, and underneath this a bed of blue skerry or bine, and this continued to the pond.

At the commencement of this drain, and during its progress, it was really astonishing to hear the ridiculous remarks made upon it by respectable farmers and others in the neighbourhood, and how much censure was bestowed upon me for cutting it. Before it was finished, I was obliged to leave that part of the country, and the work was completed under the direction of Mr. R. Parkinson, (the gentleman alluded to in the former part of this work,) and who had previously been called in by me to inspect it, and report upon it to my employers, in answer to the remarks made during the time it was proceeding. I have not had an opportunity of seeing it since its completion, but the use and utility of it I have understood to be very great; and the following extract from a letter by a gentleman, who has already written a great deal upon "Draining, and Subsoil Ploughing," &c. &c., and who has bestowed an immense deal of time, and made pecuniary sacrifices, in order to promote the cause of Agriculture, and to stimulate the tillers of the soil to make still further progress, and exert themselves more fully in the drainage of the land, will form, in a few words, the best criterion by which to judge of its value:—

"Collingham, Feb. 3, 1843.
"Dear Sir,
"* * * * You will be glad to hear that "the drain which you commenced, and which "Mr. Parkinson finished, has *completely* dried "the land, and removed the scepticism of those "who doubted its utility.

"I am, dear Sir,
"Yours truly,
"J. WEST."
"Mr. H. Huchinson, Southorpe or Walcot."

Mr. West several times viewed the above drain during its progress, and had an excellent opportunity of hearing some of the remarks which were levelled, both at the projector and also the work, and particularly one, of "young men, or rather boys, teaching old men how to drain," made by more than one old gentleman in that neighbourhood.

The cuttings of the above drain started at about eighteen inches below the surface, and at the turn in the first field had reached about six feet; from thence to the second fence the ground gradually rose to ten feet six inches, and from thence to the end about seventeen feet; the cost of it would be from £50 to £60.

The next drain I shall refer to is one cut in 1842 and 1843, between the estates of the

Marquis of Exeter and Henry Nevile, Esq., in the parish of Southorpe, in the county of Northampton. This parish has lately undergone an enclosure, and the land in question is a portion belonging to each of the above proprietors, commencing on Mr. Nevile's estate, running into that of Lord Exeter, and terminating upon that belonging to Mr. Nevile. The piece of land in question was one continued line of bog and rushes, with apparently three large outbreaks for the water, which covered the ground nearly all over; a portion of it making its escape through an old tunnel, across a field, into a fence dyke, running through the village. To effect the drainage of this piece, I started from where the dyke terminated, through the fence, into the adjoining field, and from thence I proceeded to surround the water, by cutting immediately above the outbreaks, as shown in the accompanying sketch, and continued it to the end. Having completed the first line of drainage as shown, the top springs were completely dried; but I found, to my astonishment, that although there had been the full quantity of bore holes put down in the drain, and everything done to secure the whole of the spring water, that I had to contend with a second body; and that although one set of springs was cut off, there was another left, before the piece could be perfectly dried. I therefore cut another line of drain lower down, to get the second springs, and this was not even then accomplished, (nor ever would have been,) without *boring*. On arriving at the second outbreak, the rods were put down to twenty-two feet, the depth of the drain being six feet, and the water laying sixteen feet below the level of the drain. The drain was continued forward for the length of six chains and a half, and three more bore holes were put in, the second being thirty-six feet, the third twenty-two feet, and the fourth twenty-six feet; but the water rose at the same level in the whole of them. The borings were through strong blue clay, and the water rose from a mixed measure of sand and gravel, at the depths stated. The object of sinking two of them below the rise of the water, or the open measure from whence it ascended, was to ascertain if any more lay below this.

The cuttings in the top drain, from the fourth chain, were through one continued line of red and white sand, sand rock, and running sand, with the exception of one place, shewing a vein of about a foot thick of exceedingly dark (between black and purple) coloured clay, similar to what is found in the coal districts of Derbyshire, and at the top end was a chain or two of gravel underneath the sand.

The cutting of the bottom drain for the first chain was through shale stone, and partly in the next chain; the rest of it through blocks of strong blue clay, red sand, yellow sand, and occasionally gravel. The drain was carried upon the level of the water, from about the second chain in the first drain, and there is a slight inclination in these two chains. Annexed I give a statement of the cuttings, shewing the average of each chain, and the quantity of hack-work where necessary to use the pickaxe :—

No. of chains.	Depth of cutting.		Hackwork.		Shovel and Toool-work.	
	Ft.	In.	Ft.	In.	Ft.	In.
1	4	0	2	0	2	0
2	6	0	5	0	1	0
3	6	0	5	0	1	0
4	7	0	5	0	2	0
5	8	0	3	0	5	0
6	9	0	3	0	6	0
7	10	6	5	0	5	6
8	11	0	5	0	6	0
9	10	0	4	0	6	0
10	10	0	3	0	7	0
11	10	0	3	0	7	0
12	11	0	3	0	8	0
13	11	0	3	0	8	0
14	11	0	3	0	8	0
15	10	6	2	0	8	6
16	10	0	4	0	6	0
17	10	0	0	0	10	0
18	10	6	0	0	10	6
19	11	0	0	0	11	0
20	11	0	3	0	8	0
21	11	0	7	0	4	0
22	10	0	8	0	2	0
23	10	0	0	0	10	0

The above will give the reader some idea of the variation from the inequality of the ground, of the cuttings from the level line, and also shew the extra labour in getting through the work; but even this drain, compared with others, is very good, as sometimes for a chain together the whole is one continued piece of hack work. The great evil in this drain was the difficulty in getting the sides to stand through the sand, even when all the necessary precautions had been taken by planking and stretching, the ground in some places giving way for ten yards long by five yards wide; and I will here undertake to assert, that no set of men upon earth, but those accustomed to it, would have ventured into these places, and none other than a practical man would have been able to put in the tiles. The drains were completed, and answered admirably; part of the land, once a bog, has been sown with oats, and part with turnips, and you may now get over any part of it. The cost of cutting the above drains, including the tiles, &c., was about eighty pounds.

The quantity of water discharged is about 30 gallons per minute, or 43,200 gallons per day, of the best and clearest spring water possible, and would be sufficient to supply the town of Boston, Lincolnshire, with fresh water.

The cutting in several parts of the drain, where there was little rock, shews the following as the result:—

	Ft.	In.
Soil	3	0
Red Sand	3	0
Red Skerry	0	6
White Sand	2	6
Strong White Sand..	1	0
	10	0

From this depth, to the lower level of the spring water in the large drain, was nine feet to the water, then one foot of yellow skerry and sand, and then eleven feet of blue clay. At another trial made, was eleven feet, as before, and seven feet of mixed sand and gravel to the yellow skerry, and below that nineteen feet of strong blue clay or blue bine. Within a quarter of a mile from this spot, or rather less, has been got the strong pendle and other stones for building, shewing how much the geological formation of this part of the country varies, and how difficult it would be for any one to describe the various strata, without having local knowledge of it.

I shall allude to one more drain in this parish, part done with tiles and part with covered stone, with stone sides, (the reason for my using stone, being the want of tiles in the neighbourhood). This drain is at the entrance to the village of Southorpe, and the land dried forms part of the stackyard of Henry Nevile, Esq.; and the stackyard wall and the footpath from Southorpe to

Barnack, passes over part of what was formerly *three* ponds. The drain commences at the ditch, which sends this water from the village into the "Welland," and by the side of the field where the waters in this parish divide; one set of springs running into the "Welland" and the other into the "Nene."

The ponds were a great nuisance, diverting the road out of its proper course, and keeping the yard constantly wet, from the springs which broke out and ran into them. I carried a drain under the new road and above the ponds, at a distance of seven or eight yards, being as high as I could get for the buildings, and cut off the whole of the supplies. In cutting the drain, a natural bore hole or outlet for the water was discovered, about half the distance, and when first cut through, the water from it rose to the height of about four inches above the level of the drain, and after a time merely a little above the water, resembling the boiling of a pot. I could not ascertain the quantity discharged from this hole, but the size of it was about two inches in diameter. The quantity of water discharged from the drain, is about fourteen inches wide and two inches deep, and this supply is constant, forming quite a stream.

There are several other deep drains in this parish, equally beneficial with the last, and proving their utility: one of these discharges

14,400 gallons per day, and another 30,240 gallons per day, being more than double the quantity; discharging a sufficiency of water to destroy the herbage and cropping of a whole lordship. Thus, places which never in the memory of man, grew any thing but rubbish, have been brought into a state of cultivation; a great portion of which had been attempted to be drained; some have completely beaten the occupiers, never having been drained at all, clearly establishing the fact, that land cannot be drained by "Shallow Draining," when subject to deep water, and that nothing but "Deep Draining" and "Boring," would have ever made this land available for the purposes of husbandry.

BASTARD DRAINING.

WHERE REQUIRED.—PIPING, HOW PERFORMED.—DRAIN EXECUTED IN NOTTINGHAMSHIRE.—DESCRIPTION OF PIPE TILES, WITH THEIR USE AND APPLICATION.—REMARKS AS TO THEIR USE IN SHALLOW DRAINING.—AND THE TIME SHEWN WHEN FIRST USED.

To some, the application of this word may appear strange, as the term is quite novel, and first had its origin amongst Drainers, or working Drainers, and is applied to that description of land which requires Draining not too deep, and yet to be deeper than Shallow Draining, varying from two feet six inches to six feet. This is the sort of draining which is generally requisite upon low table land, or flat land; and I have seen very few instances where it has been required upon high land. The land generally consists of either a brown or black soil, sometimes black or red sand soil, with sand and gravel underneath, and very frequently the subsoil is intermixed with layers or strata of clay, sometimes blue, but more frequently yellow. In cutting drains upon this description of land, a great deal will depend upon the outfall, it being almost impossible to get beyond a certain depth, say the mouth of the outfall, not more than eighteen inches; but owing to the dip of the land, it will enable you, by keeping the level, to go at least five feet at the top end. This I have endeavoured to shew in the five acre field drained upon this principle, marked No. 2; but there are many fields which require this sort of draining, where you cannot get beyond three or four feet deep. The size of the field must direct and guide you as to the number of drains you will find it necessary to put in; and a careful examination in the first instance, will tell you how far the water will draw, as, should the water rise at the head of the land, one side and one head drain will be all that is requisite; whereas, should the water break in in several places, it will be necessary to have one outfall drain, and the leading drains into it, at the distance of two chains, or perhaps even three chains will be near enough.

One of the first steps in this, as in " Deep Draining," is to make holes in the piece upon the dry land, and ascertain at what depth the water lays, and how high it will rise, and as soon as any one of your drains is cut, you will ascertain when it reaches the neighbourhood, or immediately opposite, or within a line of one of the trial holes, whether the water draws from the hole into the drain or not. It will be said the water lays so deep in this land that it cannot be got into the level of the drain; this is a mistaken notion, and one which has not only baffled many in draining, but is only understood and cannot be executed by any but a practical workman. The drain is carried forward at the water level, until it reaches a soft porous place, where the water purges up above the bottom; here it will be necessary to make a cutting from the line of drain, about six feet square, (and more, if the ground will not stand,) and having dug this out in the centre, three or four feet below the level of the drain, and if the water should rise very rapidly, it will also be necessary to keep two men to bale it out. The Drainer then proceeds to put in some loose hay, or long dry grass, together with some stones, so as to prevent the sand from rising too much; he then places two tiles perpendicular upon this, the edges standing against each other, and effectually secures them all round with clay; having completed these, he places two more on the top of these tiles, and so on to bring them to the level of the water line in the drain; the water then, from the pressure underneath, ascends to the top of the tiles, and passes down the drain with the rest of the water.

This is termed "Piping" a drain, and is, next to "Boring," one of the most important steps connected with the art; as, if this water which is secured by piping, were allowed to remain in the land, and could not be secured by the drain, the drainage would most certainly be incomplete. There are other modes of doing this, but this is one of the most simple and easy to accomplish. One plan would be by bricking and cementing a small well in the place; this would be attended with a great deal of trouble, and also expense: another would be, after securing the bottom, which must be done in both cases, against sand and silt, to let down a barrel, not too wide, after the pattern of a herring cask, and bringing the water up it to the level of the drain. This has been tried and found to answer in places where great difficulty has arisen in getting rid of the water rising from the bottom, as it could be readily placed in its position, and the earth around it filled in afterwards.

In draining the water meadows belonging to the Earl Manvers, at Palethorpe, Nottinghamshire, a few years ago, or rather, I should say, in completing the drainage, (as the original drain was cut by Tebbet, about twenty years ago, and carried under the bed of the river into an open drain that was brought from a lower level;) the tile used by Tebbet was six inches by seven inches, and not doing its work effectually—the water not passing off rapidly enough—the land shewing young rushes to a considerable extent, and no blame attributed to the original Drainer, but the want of a larger tile; a well was made in the small meadow to the right of the road, to the Kennels, about fourteen feet deep, and a box curb, six feet square, was let down and made secure, to the height or level of the tile drain, which was eight feet deep: the water then passed from the well down the drain into a tile nine inches by ten inches, being the largest sized tile made. An additional depth or fall was obtained in bringing the drain up again, and, in addition to the tile nine inches by ten inches, a tile six inches by seven inches, (being the first tile put in,) was run up alongside of it, and this was nearly filled as well as the larger one.

Reference is made to this drain more particularly, from its having to be piped, or a well made, its proper place being under the head of Deep Draining; yet, lying as it does in table or flat land, it is only from its extreme length, being near upon half a mile, and its depth at the well eight feet, makes it approach Deep Draining, the fall being obtained to get lower, by carrying it under the bed of the river.

During the execution of the above work, and to get the well completed, it was necessary that the water should be kept out of the way of the workmen: and after trying the aid of a pump, it was found requisite to have a bucket containing sixteen gallons, to draw up the water. The ground adjoining was principally running sand, but the influx of water in the bottom was so great, that the quantity drawn out would be about fifty gallons of water to one gallon of sand. This was a very important job, and reflected great credit upon Samuel Smith, the Drainer, who, at that time, executed the works of Earl Manvers; in fact, it could not have been executed by any one but a practised Drainer, the difficulties to contend with, from the earth giving way, as well as the body of water, requiring great skill and perseverance. I have not seen this drain since it was finished, to ascertain whether the end desired was obtained, but, from authentic sources, I understand it was quite satisfactory.

It will sometimes happen in cutting these drains, that places in the bottom will be found so soft, as not to allow the flat tile to lay on its proper level;—in such cases, it will require the soft mud to be thrown out, and if stones are near which can be used, flat enough for the purpose, or tile pieces if you are near enough to a brick-yard, to spread on the bottom, so as to raise the flat tile to its proper level. But, here again, if your foreman or drainer has not been accustomed to laying tiles, he is as liable to get it two inches too high as two inches too low, and therefore care is required in keeping the drain at its proper level throughout.

It will be necessary to put in the tiles in these places with refuse hay or long dry grass, at the bottom of the soft places, and for the tile to be well covered with sods, if possible, and then for some soil to be put on the top of the sod, after which, all the open stuff (that is, the sand or gravel,) may be thrown in and the drain filled up. Any clay which may be dug out should be spread over the land, the same as I have recommended in Shallow Draining, and it will incorporate with the surface soil, and be of as much benefit to it as a good dressing of manure.

The tile required for this description of Draining generally, for the outfall, should be five inches by six inches, and for the rest of the drains, if many are required, four inches by five inches; but if only one drain is requisite, the five inches by six inches should be used throughout; and unless the bottom be strong hard gravel, it will be necessary to use flat tiles for the whole length.

There is another description of tile very useful in this sort of Draining, which has been but very little used, or the use of it very little known, which may be applied in piping the drain, or in carrying the level of the tile in the soft parts of the drain, and this is called the pipe tile. This tile is made with a shoulder on the outside of the lesser end, and a groove on the inside of the larger end, so that the smaller end will pass into the larger one to the extent of one and a half or two inches, in some tiles more, and these tiles, when well clayed, will carry water for any distance. They are also made without any shoulder at all, and fit with the smaller end into the larger one, and are in shape like the chimney pot, but less in diameter.

This description of tile is of the greatest use in carrying a drain across a running sand, where you do not require to catch any water in the bottom, and, at the same time, wish to carry your water forward into some of the open drains near; it is also of use in carrying water either under or over other drains and water courses, where you intend to get an additional fall, or to carry your water out at a lower level.

These tiles are coming more generally into use than formerly, or they were some years ago, and are intended, in the present day, for the purposes of Shallow Draining. Having omitted noticing them in their place as connected with Shallow Draining, and having shewn to what purposes they are applicable, I shall here endeavour to give my ideas, as connected with the object to which they are now applied.

In the process of Shallow and other Draining, the bottom of the drain requires to be perfectly level and sound, that is, so as not only to bear the tile in its proper position, when placed in the ground, but also, if necessary, to keep the same inclination when using flat tiles, and in the use of these, half the tile rests upon one flat and the other half on the next flat, giving stability to the drain in the same manner that the bricklayer gives stability to a wall by tying one over the other: such being the case, the tiles are less likely to be disturbed or give way. One great feature in the new tile is, that you are said not to require flats, but that the cylindrical or pipe tile answers the purpose both of arches and flats; but we have yet to learn whether with the same success or not.

I have no wish to discourage Drainage in any form, but, as before stated, "whatever you do, do well;" and, therefore, I must remark upon these pipe tiles, having all along recommended "Tile Draining." This new attempt is making, or causing a mutiny in the camp. If your article is, not suited to the purpose to which it is applied, it is hardly likely to answer; and there are very great doubts whether these tiles will or not, except in particular soils.

I have not seen these tiles used for Shallow Draining, but unless great care is taken in cutting the drain, and laying them in the ground, they will be likely to shift from their places: this will also happen if the bottom is soft, as they will be likely to sink, and the soil following, the drain will be made up. In a statement made to me, they are said to be fourteen inches long, and from three and a half to one and a half inches in the opening, varying in size; these are laid in the ground with the ends to each other, and have therefore, nothing but the subsoil to rest upon. As I have before stated, the action of the atmosphere upon the clay will, before the tiles are put into the ground, render it so soft as to be necessary to use the flats for the tiles to rest upon; therefore, if these tiles are put in without support—and such is their object—the pressure upon the surface, in a few years, in this description of land, will remove them out of their places, and then the old system of taking them up, and replacing them, will have to be adopted. Those, therefore, who do use them for the purposes of Shallow Draining, should be careful to see that they are not only laid in the ground true, but that they are not used in soft places, where they are likely to give way. I know it will be argued that they have answered, that they do answer; but what I would impress upon my readers is, not only to think of the present, but to consider the future.

There is another point too, connected with them, as compared with the other tiles: they are not calculated to draw off the water so rapidly. The tiles I recommend not only standing higher, but the interstices of the sole or flat, and the arch tile together, admitting of the water passing in more freely than it can into the pipes. A very great evil or defect in draining has been in having tiles too small.

There is a larger sized pipe tile, about six inches in diameter, made for the purpose of carrying water in and about buildings, or in the conveyance of large supplies from one place to another: these are made about a foot or fourteen inches long, with shoulder and groove, and will turn round buildings, or other places, where required.

The manufacture of pipe tiles, to aid in drainage, may be traced back for several years, and yet at the present time we have patents started for their manufacture, as if they were quite new; and amongst those who tried them as far back as thirty years ago, I may mention Russell Collet, Esq., of the Jungle, Swinethorpe, Lincolnshire. Whilst speaking of this eminent "Old English Gentleman," and practical Agriculturalist, who was amongst the early drainers of land, I may here state, that he had done an immense quantity on the estate where he resides, a portion of which is strong clay, and more of which is of the description requiring Bastard Draining, stated in the preceding pages. I had the pleasure of riding over his farm in 1841, and although he is far advanced in years, I saw extensive alterations about to take place, both in the fences, the draining, and the ditches. Upon inquiring of him the cause of all this, he stated that he had just discovered that a very large portion of his draining required doing over again, for he found, upon wishing to subsoil the land, that scarcely any of the tiles were put in sufficiently deep to be out of the way of the subsoil plough, and finding this to be the case, he had resolved to do it all over again, bearing in mind, that if he did not live to derive the benefit of it, those who succeeded him would.

This forms another instance where practical men have erred in the performance of Draining, during the last half century. Even in the common course of draining land, as upon the estate above alluded to, two distinct modes of draining were requisite, nevertheless years are allowed to pass over before the parties discovered their error, who, with the best intentions possible, have tried draining.

The process of Bastard Draining is equally simple in its working as in Shallow Draining, and if parties, having the description of land pointed out, will adopt the advice here given, they will be enabled to dry their land, not only thoroughly, but securely and effectually.

IMPEDIMENTS TO DRAINING.

DRAINS ARE LIABLE TO BE STOPPED UP.—STOPPAGE OF A DEEP DRAIN AT WELBECK GARDENS, NOTTINGHAMSIRE, BY HORSE RADISH ROOTS.—THE LIKE IN EDWINSTOWE MEADOWS BY THE ROOTS OF GORSE.—THE LIKE AT SAUCETHORPE, LINCOLNSHIRE, BY THE ROOTS OF AN ELM.—THE LIKE AT THORESBY, NOTTINGHAMSHIRE, BY THE ROOTS OF A WILLOW.—THE PROCESS OF SCOURING A DRAIN.—THE STOPPAGE OF OTHER DRAINS, AND THE CAUSES.—STOPPAGE OF A DRAIN BY THE ROOTS OF AN OAK.—WATER WILL SHALE IN DRAINS, WITH HINTS FOR PREVENTION.

HAVING given a description of a few of the leading Drains which I have seen executed, within the last fourteen or fifteen years, and the result of them, there is one important part of the work yet to be explained, and which very frequently has (for a time) destroyed all the good which the Drainer has effected, for want of a more timely acquaintance or knowledge of the evil on his part;—this has happened in a variety of instances, and has seriously injured the work. What I here allude to, is carrying deep drains too near to fences, or in the neighbourhood of trees, which have, from the removal of the soil in their vicinity, been able, from

extending their roots in support of the parent stem, to work their fibres into this shifted earth, and gradually into the drain, and so completely stop it up, thus preventing the water from passing along it, and thereby rendering the drain of no effect beyond the part where such stoppage has taken place. I shall relate several instances which have occurred, that I have had practical knowledge of, and which I trust will serve as a guide to those who wish to execute draining properly. One or two instances of this sort have already been mentioned to the Royal Agricultural Society; but as they were in part copied for those who made use of them, by myself, I shall here name them again, as in part connected with this work. The first instance of the kind is related by J. E. Denison, Esq., M.P., of Ossington Hall, Nottinghamshire, who, in his report on the water meadows at Clipstone, says, "Too much
" care cannot be taken never to carry these
" drains within a very considerable distance of
" trees; their roots seem to be attracted in a
" wonderful manner by the moisture of the
" drains, and if they once find their way into the
" tiles, they throw out bunches of fibres, which
" soon mat together and stop the drains, and it
" is astonishing the depth that roots will descend
" after the water, apparently in search of it. A
" deep drain on the outside of the garden wall,
" at Welbeck Abbey, (the seat of the Duke of

"Portland,) was stopped at the depth of seven feet in the ground, by the roots of some horse radish plants, which had found their way into it."

The next instance occurred under my own observation, and of which I had "ocular demonstration." A drain was cut in a piece of land in the parish of Edwinstowe, above the level of the water meadows, adjoining to the township of Clipstone; it was intended for a soke or side drain, and a large tile was put into it; at a distance of fourteen feet from the drain grew a quantity of strong gorse, the roots of which ran along down a sloping bank, and into the drain, and choked it completely.

The next was a drain cut in the parish of Saucethorpe, in Lincolnshire, on the other side of Horncastle, and near to Spilsby, upon an estate belonging to the Rev. Francis Swan, (I do not recollect whether this drain was cut by Tebbet, or whether by Samuel Smith, but I am inclined to think by the latter—at all events it was opened by Smith). The drain was four feet deep, and at a distance of fifty yards from it grew a lot of elm trees; between these and the drain ran a sod bank, dividing a piece of grass land from a piece of arable. The root from one of these elms ran along the sod bank, which, from its soft and vegetative quality, was no doubt assisted very much in its growth, and after travelling this distance, found its way into the drain, which, after a time, it completely filled. When the drain was executed, the land was to all appearance, perfectly cured, and from its wet state a few years after, it was deemed necessary to open it again, when, to the surprise of those who opened it, it was most completely filled with the root. The length of time it was growing cannot be ascertained, nor does it appear that the elm was attracted by the water, having already found its way into the sod bank, from the removal of the soil in cutting the drain; the loose mode of filling the drain up again, may have caused the root, in obedience to the law of nature, to again descend into the earth, instead of continuing its course along the sod bank; it would have done this had the drain not been cut, and thus having found its way to the water, of which elms are particularly partial, it grew so rapidly as to fill up the drain.

A drain cut to dry a flat of land in Thoresby Park, Nottinghamshire, was found, after several years, to decrease in its usual stream, and to shew, by the water on the top, that the drain was either made up, or else that the water had found another outlet, and ceased to dry the land effectually. It was opened and cleaned out, at intervals of every ten or from that to twenty yards, and was completely scoured or cleansed for a quarter of a mile, when it was discovered, at a turn near the head of it, that the roots of a willow tree, several yards from it, had found their way into the tile, and completely stopped it, so that all the water above this part was forced on the top of the land, and covered it. The root, when taken out, was similar to the root of an old plant taken out of a flower-pot, where it has been confined for many years, but in size or length considerably larger, and as fast in the tile as if it had been rammed or plugged into it.

The process of cleansing or scouring is carried on in the following manner:—The drain is opened at intervals, sufficient to put down a tiling lath, with a small wisp upon it, which is passed along the drain backwards and forwards, so as to completely remove anything in the tile; or in lieu of this, if there is sufficient water, to float a piece of cork with fine twine down the drain, and then attach a strong line and draw it after, with a wisp of hay tied in the centre, and worked backwards and forwards as before, until the tile is thoroughly scoured out, and this is repeated unto the end.

A drain cut upon a farm belonging to the Earl of Scarborough, in the occupation of Mr. Flower, at Bilsthorpe, Nottinghamshire, which was only two feet deep, was made up with the roots of a weed called pig weed; and in the same parish it was found that the roots of nettles, allowed to grow thick over shallow drains, had filled them up; and there is also an instance of an elm tree in this parish crossing a gravel road, under the formation of the road, into a drain two feet deep, and filling it up.

The last I shall mention was a deep drain in the Cockglode Meadows, above the town of Ollerton, Nottinghamshire, cut some years ago, and opened a few years since. Above this drain grew some large oak trees, at a distance of about ten yards; the drain was about eight feet deep, and the roots from one of the oaks found its way into it, and filled it up; the form of the root was in appearance a good deal like a hedge hog. The same reason may be assigned here as for the elm root at Saucethorpe. These trees were from two to three hundred years old, and perhaps more, and in all probability when the drain was cut, one of the old roots started a fresh shoot in the newly opened land, and then found its way into the drain, and being assisted by the air and water, thus grew rapidly, and choked it up.

There is another remark too, which must not be omitted here, and that is, in some drains water will shale or form an incrustation upon the flat tile, and in a few years the drain will require cleansing: the form of it and appearance will be a good deal like what is called fir in a tea kettle from long usage. I am not sufficiently acquainted with chymistry to point out the quality of the water where this formation takes place, and must therefore leave it to parties to draw their own conclusions.

By the foregoing remarks upon the liability of roots to fill up drains, and also of this formation from water itself; I cannot too strongly point out the necessity of avoiding these obstructions in the first instance, and by keeping perfectly clear of trees, to run no risk of having them stopt up, and also to prevent their roots (as well as those of other plants that search deep into the earth for support) from getting into the deep drains, by having the tiles well covered with sods, and well filled up with the soil or other strata, so as to make the land secure after cutting. The whole of what is thrown out in the deep drains may be cast in again, as well as a large portion of it raised up over the drain to permit of its settling down again; for even if it is rammed, no injury can arise to the drain, as the tile is the line of water, and it is in the neighbourhood thereof, that the great body of water makes its appearance. No portion of it is intended to take in top water, and therefore is it unnecessary to keep it loose or open, unless there should be any outbreak above the line of drain, and this can be removed by putting in loose stones or tile pieces, to let it into the drain over the top of the sod, which allows it to filter through. I have seen parties who have set out drains, and carried them forward with a good working man, who have had a large quantity of stones placed in over the top of the tiles for a foot or two, but this is perfectly unnecessary, and an absolute waste of labour and material. Everything should be done to make the land secure after the completion of the drain; and I would recommend to parties, by careful attention and observation from time to time, to watch their drains, so as to observe that no obstruction to the free course of the water takes place, and, when it is observed, by timely and judicious treatment, prevent the evil from becoming greater. It is quite evident that with draining as with other things, it is not always perfect, (or rather, it has not as yet been so,) and from the foregoing remarks, the reader will see that it is possible to blame a man for not having done his work effectually, when, it is very apparent, that the causes of obstruction to perfect drainage, have not even come within his knowledge; and that, however good his work might be at the time, and for a a length of time afterwards, it has been damaged and injured by causes which he never for a moment contemplated. In parishes or townships where a great quantity of draining has been done, and also upon the estates of Noblemen and Gentlemen, if the party who performed the work is near, it would be advisable to let him examine such drains once or twice a year, to see that the mouths discharge properly—that the land is dry, and the drains performing their work—and, if any little matter wants correcting, to let him do it, or see that it is done: the expense of this will be very trifling; and it is these little attentions which very often save parties a great deal of labour and expense, when if neglected or unattended to, they become very serious.

BORING.

ITS ORIGIN.—ARTESIAN WELLS, CONNECTED WITH IT—BORING FOR, IN VARIOUS PLACES IN FRANCE, AND THE RESULT.—DESCRIPTION OF BORING RODS.—BORE HOLES IN NOTTINGHAMSHIRE.—WELLS DRIED BY BORING.—BORE HOLES IN NORTHAMPTONSHIRE.—GENERAL REMARKS UPON.—AND THE VALUE OF TANKS TO FARMS FOR TAKING THE RAIN WATER.

THE usefulness and utility of Boring have been but very little considered throughout the kingdom. By some its advantages have never been considered at all, nor yet has it entered into the minds of many who have performed " Deep Draining," that to make their work blameless and free from censure, it would be necessary to have recourse to Boring at all. That it is both requisite and necessary, in many instances, I shall here endeavour to point out.

The origin of Boring is somewhat obscure, but the art of sinking Wells, to which it is in a great measure allied, has been known to exist for ages—even back nearly to the flood. It is not here necessary, however, to travel back for any great period, although it may perhaps be a matter of doubt, whether the ancients had not arrived at the same methods of discovering water, or whether, after slumbering for years, we may not be using the very same implements (but of a better quality) for effecting the object of tapping the water, which they used: certain it is, that in our own nation, wild and uncultivated as we were for ages, nothing of the kind could have reached us until within the last few centuries, although very wonderful discoveries have been made by talented individuals during that period. But if the discovery of Boring, as well as other works connected with the conveyance of water, had not made any degree of progress in this country, it is evident that the Romans, in their glorious days, had acquired considerable proficiency in the art, not only of raising but also of carrying water to very considerable distances, the remains of their fountains and aqueducts existing even at the present day.

That Wells and Boring are intimately connected, is more fully shewn, I think, in the present day, from the term "Artesian" being now applied to certain Wells, which are, in fact, "Bore-holes." The derivation of this term is from the French province of Artois, where no water can be obtained but by boring to considerable depths; and, from some of the articles published within the last few years, I shall here give a statement of a few of these Wells, as connected with Boring, and afterwards shew the effect of some made in our country, connected with Draining.

These wells (and in some instances fountains) are formed by perforating the earth with rods, until a subterraneous body of water is tapped, the sources or means of supply of which are higher than the spot where the operation takes place; and when once the spring is tapped, it immediately ascends to the surface; or, if the pressure be great, it will rise in proportion to the weight pressing upon it. I have before attempted to account for the rise of water, which causes "Deep Draining" to be necessary upon land, and I perhaps may not improperly introduce here, in connexion with wells, some of the hidden supplies of water passing under ground, which will account for the variation in the quantity, or volume of water uttered by these wells; and to illustrate this more fully, it may be stated, that after heavy rains, the mines in Cornwall shew considerable increase in the quantity of water in them, and this, not by the rain passing down the shafts of the mines, but by percolating through the different strata of the earth, until it reaches the workings in the mines. Such of my readers who have seen a section of a coal mine, or a mine for obtaining the ore, will understand this more fully.

The fountain of Nismes, in France, throws out, when at its lowest volume, about 280 gallons per minute, but if heavy rains fall in the North West, although at a distance of several miles, its volume is increased to 2,000 gallons per minute, or to the enormous quantity of 2,880,000 gallons per day. The fountain of Vauchi, in the South of France, if it were to receive all the rain which fell in a whole year on an extent of thirty square leagues, would not be supplied with the quantity it issues. Its volume of water is estimated at 480 square yards per minute, and at times it is supposed to be from 1,000 to 1,500; this is an extraordinary instance of hidden supplies, and shews that other means, besides rain, are instrumental in supplying this immense fountain.

A cavern is mentioned by Humboldt, in South America, about twenty-five yards high, and twenty-seven or twenty-eight broad, where a traveller can walk for 800 yards in length, and in whose recesses are rolled the waters of a stream ten yards wide. There are numerous other instances which might be enumerated, of the extraordinary under-ground travels of water; and in the Artesian well, at Tours, in France, the subterraneous line of communication is more fully developed. In 1831, a trial was made by boring, and a tube was inserted in the bore-hole to keep open the hole at the top, and at a depth of 110 yards, a quantity of stalks and roots of marshy plants, and seeds of various kinds, were brought to the surface, in a state which shewed that they had, from their appearance, been about four months performing their hidden voyage, and, at the same time, small shells were thrown up, similar to those to be found upon low marshy land.

An instance is related of the effect of Boring at Fontainbleau, in France. The workmen had been engaged for some time, and the progress of the work was slow, when all of a sudden, the boring rods descended for about eight yards, and great difficulty was experienced in endeavouring to draw them up. In consequence of the suction of the water, which they had reached, this is no unusual occurrence, and instances have been known of attempts at Boring, where, in consequence of suction or adhesion, the rods have not been got out at all, and again, where they have been recovered, after an immense sacrifice of labour.

Amongst other wells, there is one at Dieppe, which is three hundred and forty yards below the surface. A well at Perpignan, which discharges about four hundred and twenty gallons per minute; and one at Tours, which rises two yards above the surface, and discharges two hundred and twenty gallons per minute. Numerous other instances of boring might be mentioned, which have taken place, not only in France, but in this kingdom, for the purpose of supplying places with water, or rather, water of a given temperature, (some of which have been requisite from having none in a given distance, and the other has been obtained for medicinal purposes). During the last few years, very important additions have been made to this department of the art, the depth attained having not only been greater, but the facilities of doing it and also of raising the rods have been increased; but these depths are far lower than it would be either requisite or necessary to assist in the process of Draining. The object of attaining it, as connected with the latter, is for the purpose of giving the water an artificial outlet, instead of a natural one, and by confining it to a given place, preventing it from ascending through the open measures, where it finds its outlet; thence flowing over the land for a great distance, and injuring it in its course, until it finds a natural outfall. The first statement which we have of Boring, connected with Draining, is the accidental discovery by Elkington, by means of an iron bar, that water lay beneath the level of his drain, and that it would either be necessary to lower his drain, or, by means of these holes, raise the water to the level of the drain, and so get rid of it.

To those who have neither heard of the instrument itself, not yet of the mode of using it, it may not be improper to give a brief description of it. In the first outset of its use a rod of iron, about ten feet long and one inch square, was used, with the bottom end of it made in the form of a carpenter's auger or gouge. It was afterwards discovered that this instrument was not sufficient to answer the purpose. It was then found that additional length could be got, (in fact any given length,) by having it made in pieces, one end being a screw, and the other a screw socket, so as to screw into each other. It was also suggested, in Johnson's remarks upon Elkington, that the boring rods might be worked horizontally into the earth, instead of perpendicularly, so as to tap the sides of hills. This process I have never seen tried, and if I had, should hardly be disposed to give any favourable opinion of it, as to its use. The probability is, in attempting this, that you might try five times out of six, without being able to cut off the supply of water. The other process I have seen, and can recommend it, as being highly beneficial and valuable as a discovery.

The boring rods which I have used are in length about four feet six inches, and are made of iron one and a half inches square; the first piece comprises the gouge or auger, about three inches and a half in diameter, and about eighteen inches long on the boring part; at the top end of this piece is the screw; the second rod has a screw at the top, and a socket at the bottom for the first to fit into, and so in succession for ten or twelve lengths or more, if requisite; about fifty feet being as deep as it is necessary to go, unless it is where difficult to obtain water. But the carriage of them from place to place is an object, and it rarely occurs that you require more than this length. In addition to the rods named, there is also a first length, in the form of a chisel, about three inches and a half in diameter, for the purpose of cutting through stone or plaster-beds, where the gouge or auger will not work; and as the stone is sometimes very hard, and requires immense labour to cut through it, unless the chisel is of the best metal possible, like other implements, it is liable to break; and it is, therefore, better to have two or more chisels. The size of both the gouge or auger, and the chisel, may be increased, if necessary, but both should have the same width or diameter, to admit of the chisel passing after the gouge. There are also two handles, and sometimes three, the first made short, about two feet long, and the next about three feet, with iron sockets to drop into the square of the handles, and fasten the rods firmly, whilst in work. There are also three or four wrenches to fix on the rods, to assist in their progress, either upwards or downwards, to draw or lift them when coming up, and to check their descent in going down. In Boring in a line of Deep Draining, the bore holes are made on the side of the drain, the same as in Piping, so as not to impede the progress of the work; and if the water by the bore-hole rises to the level of the drain, it is then passed by a tile into it, and so flows with the drain water. Over the drain, at the time of working, is placed a stage of boards or planks, with a hole left for the rods to turn round in, and the foreman stands below at the hole by the level of the tile drain, to empty the rods, and to steady them in work, either in going up or down, as well as to remove the accumulations each time from the gouge: and he is also able to tell, from the measures brought up, (if he has been at it for some time), how much farther it may be requisite to go to reach the water, and whether to make any alterations in the work. Tebbet, during his life, was very partial to boring, and effected some very extensive improvements by means of that alone, (without reference to Draining,) in various parts of Nottinghamshire and other places; and like results have ensued to others from adopting this plan, in other parts of the country; but as the rods are not got without considerable expense, (from £15 to £20,) very few are in possession of them, nor would it be wise for any one to be at that expense, unless for draining a large quantity of land, or for the owners of large estates.

About twelve years ago, (1831,) a bore-hole was tried in the hamlet of Moorhouse, Nottinghamshire, on land in the occupation of Mr. Samuel Pinder; this hole was thirty-six feet deep. The first portion of soil was a mixture of red soil and clay together, afterwards black peat soil, and then red clay and skerry or bine; the cutting through these was about thirty-four feet, and then a strong measure or strata of stone, twenty inches in thickness. The first portion of the stone took four men one hour to cut an inch of it, and they completed the whole of the twenty inches in five hours; and as soon as the stone was perforated, the water came up to the surface of the land, and afterwards went into a ditch adjoining. The first discharge was at the rate of one hundred and twenty gallons per minute, and it has continued to discharge at the rate of one hundred gallons per minute for a considerable portion of the time since. I am not prepared to state what is the volume of water now, but have understood it to be about as usual. Another hole was made about the same time, in the same parish, lower down. This was bored to the depth of forty-five feet, and the average run of water from this is about forty gallons per minute. The effect of these holes was not apparent for some considerable time afterwards, nor is it to be expected that land, which has been saturated with water for years, can at once be got perfectly dry; but these holes have made the land dry to a very considerable distance; and although much has been done in Shallow Draining the land there, the greatest improvement with these holes was, that they tapped the supply of spring water, flowing or breaking out upon the lands in the neighbourhood.

The next to which I shall allude was the drain before-mentioned in this work, at Kneesal,* Not-

* I should think the benefits of Draining were never so fully and clearly shown, in any parish in the kingdom, as in this. In the year 1829, I first became acquainted with it, and with the occupiers; and at that time it was considered the wettest and coldest strong-land parish in that part of the country. The land was poor, the tenants were poor, and, to add to the whole, they were high rented. There was scarcely a man in the parish who could pay his rent when it became due. Some alterations took place in the land, and the occupation of it was made better for the tenantry. At this time there was a brick-yard upon the estate, at which a large quantity of draining tiles were made, from one to two hundred thousand per year; and these were given to the tenants, for leading, but then they had not the means of putting them in; what was done, therefore, was principally at the landlord's expense, and it was no unusual thing to see a large stock of tiles in the yard at the beginning of another season. The Deep Draining requisite in this parish was all at the landlord's expense. In the year 1833, an alteration was made by the late J. S. Bayldon and W. J. Pickin, Esqrs., in the rental, by putting the rents at what it was supposed each tenant could afford to pay, and reducing the whole about 20 per cent.

The drainage of the parish still kept progressing, and an additional quantity of tiles each year were taken from the yard. The benefits and advantages of Draining were also becoming every day more apparent to those who got some of it done, and more particularly to those who had done very little. About 1836 a decided improvement began to appear in the parish, the line of turnpike road through it having

tinghamshire, (within two miles, as the crow flies, of the hamlet of Moorhouse,) on a farm occupied by William Townrowe, which, from being one of the worst in the lordship, afterwards became equally productive with the rest. The drain was cut at an average depth of six feet, and a very considerable supply of water obtained; it was intended to place a trough in the field to receive the water, and make a watering-place for the stock, for either that or the adjoining field,

made great alterations, removing old houses, and new ones springing up in their stead, and other old ones repaired. A change in one or two of the tenants was attended with very great good. In 1838 and 1839, still further improvements had been effected, and a great portion of the draining completed. The fields had begun to assume a different appearance, the quantity and quality of corn grown was both greater and better; (the times, it is true, had been rather better for two or three years). No tiles were left in the yard at the end of the season; and in the space of ten years, this parish, from being one of the worst in the county, had assumed an appearance equal to its neighbours; and instead of all being in arrears, there was but one farmer in the parish who was not able to pay, and he partly from inattention, and partly from causes over which he had no control; thus showing what great good may be effected in a few years. The reduction of the rent I consider one of the least considerations, but it was done in the face of bad times, and was certainly a boon to the tenants. There is now no cause of complaint amongst them, and from being excessively poor, they have now become contented and happy.

and supply both closes when stocked; but on putting down the boring rods, in the place where the trough was to have been fixed, to try the nature of the ground, about twenty-two feet deep, the water from the drain passed into the bore-hole and disappeared, and still continued as fast as the drain discharged it; and this water again made its appearance about two hundred yards lower down, in a large pond supplied by a spring never known to be dry. The boring rods were then put down on the top side of this pond, until the water ascended to the surface, and then went over into the fence ditch; the soil above the pond was a strong retentive clay, and where the bore-hole was twenty-six feet deep. After this the rods were put down into the pond, and after passing them through a bed of plaster eight feet thick, they were drawn out; immediately afterwards, or as soon after as possible, the pond and drain water altogether disappeared, drying not only the pond, but the whole of the land to a very considerable extent around it, and making dry land where it used to be almost impossible to walk over at times, much less to ride. A similar result occurred upon one of the farms adjoining, in the occupation of Mr. Richard Cougill. A spring upon this farm was never known to be dry, and by boring, (not for the purpose of drying the spring, but the land,) it was rendered perfectly so. I should also state, that by this boring, a well, full three quarters of a mile distant, was drawn off by this hole. This well was one which was a great deal of trouble, as well as expense, on the first sinking of it; it was fifteen yards deep, but from the drainage in the parish, it was deepened to thirty-six yards, and it was at this time that the water was drawn off by the bore-hole in the land just named: it was then bored lower, but upon the water rising to the open measure or plaster bed, it again disappeared in the same manner that the water disappeared from the well in the first instance. It was then sunk below this level, and bored to the depth of forty-eight yards, when a good supply of water was obtained, and still continues. The only way in which this water could have been secured in the well, would have been, by inserting in the bore-hole a pipe or tube, which would have covered that part where the open measure was, then an additional height of water might have been obtained in the well; but as this was not tried, it was necessary to lower the well itself to this level, and then bore deeper to obtain from a new source the supply required. This is not a solitary instance of wells being dried, by drawing off the water in the same line, or in the same measure or level, of that where the operation of boring takes place.

One or two instances are sufficient, and will

answer the purpose, of showing the value of boring. I shall, therefore, mention one other of my operations. In drying the ponds at Southorpe Farm, it was discovered a few weeks after, that a small well, which required about three steps to get to the water, and which had never been known to be dry, even in the driest season, was found to get less by degrees, and was at last completely dry, so as to be of no further use, and was filled up by the tenant, Mr. W. Wiles, who, from having a good supply of water, was then without any. Within thirty yards of the same place, and a little further south, he dug down to the depth of eight feet, and made a hole about five feet lower with a tool, but could not find any signs of water. The whole of the strata was like stiff mortar, rather red, and would, to all appearance, have done for the inside of stone walls. I sent the men to bore by the side of this hole, and desired them to bore twenty feet, but to go lower in case they did not find the water. They had scarcely got the rods down to twenty feet, before the water rose so fast that they could hardly get out of the hole in time to get the rods out, and stop it. The water rose to within three feet of the surface, and continues, a stone being placed over the hole, which allows it to ascend gradually, and from having to fetch it a distance from the old place, they then had a far better and purer supply very near the kitchen door, so that we may adopt the old maxim in this instance, " Out of evil comes good."

I shall not enter further into any particulars of Boring, than by again referring to the drain before mentioned in this parish, where it would have been impossible, from the nature of the land, and the extraordinary outbreaks of the water, to have effectually dried this land, without boring, as there were two distinct outbreaks of spring water, issuing through two lines or open measures in the land, and from the weak state of the sand on the side of the drain, unless it had been cut very wide in opening it, no man would have been able to put the tiles into it, and if boring could not have been adopted, or had recourse to, the drain would have been ineffectual, and the greatest portion of the money expended in drying, or attempting to dry it, would have been thrown away; and this will apply in scores of other cases as well as this, and it will not be a matter of surprise to those who have attempted " Deep Draining," in districts where requisite and where this has occurred, if without the aid here alluded to, they have failed in making their work perfect.

I have endeavoured to point out, by these few practical results of Boring, the absolute necessity there is for it as connected with Draining; and I have, to the best of my skill and ability, shewn the working of the system. I do not mean to advocate it in all cases, and therefore those who adopt it, must be guided as much as possible by the nature of the cuttings in the land they wish to drain, not to try and bore in gravel or other open measures, because there the water may be expected to rise naturally, but where it is adopted, it must be where strong blocks of clay intersect the bottom of the drain; and no water is found in passing over shale, or strong bine or skerry. It may be objected to by some, that this process will, in all probability, injure the wells in the neighbourhood, but it does not often happen that either Deep Draining or Boring is necessary close to a farmstead. To the objection of taking away the supply of water, I may say, that if a large portion of your land is to be benefitted by it, why let it be done, and the same means which you have used for taking it away, may be had recourse to for getting an additional supply, and at one-sixth of the expense of sinking the well deeper; and if the land is to be dried to make it productive, it will be advisable to make this trifling sacrifice, rather than keep the land in its filthy state, worth literally nothing, when it can, by a trifling additional outlay, be made available for producing food, sufficient to satisfy the wants of scores of families, and thereby conferring a benefit upon the community at large.

To meet the demand for water in large farms, there is one thing which has latterly attracted the attention of landlords and agents, in erecting new buildings, and as, by adopting my plans, they may be likely to run short, I would strongly impress upon all, the necessity of spouting their buildings, and by having a large tank made in some convenient spot near, so as to collect all the supplies into it, which can be made available, not only for household purposes, but also for the stock; and although the cost of such tank may be considerable in the first instance, its advantages, in both cases, will be very apparent in a few years, and an immense quantity of water prevented from filling the yards, and washing away the strength of the manure, which ought to be employed in producing future crops.

LABOUR.

EXPENSES OF, HOW REGULATED IN SHALLOW DRAINING.—DEEP DRAINING.—BASTARD DRAINING AND PIPING, AND ALSO BORING.

In the foregoing remarks, I have endeavoured to shew the labour upon each description of Draining; and I have also laid it down as a rule, that in all departments of the work, I would have it performed by measure instead of by the day; at the same time I would state, with regard to the amount to be paid for the work, it must in every case depend upon the soil through which you have to cut, to effect your drainage. No fixed sum can be named for doing the work in different localities; for whilst in one place it may be all digging, in another it may be a quarter or half of it hack-work, and the rest portion of it digging or tool-work; or, whilst one side of a field is all tool-work, the other side of it may be half tool and half hack-work; so that with this extraordinary variation in soils, it would be a very difficult matter to lay down a price to guide you in the work throughout; nevertheless, bearing in mind that the "labourer is worthy of his hire," my plan has been to inspect the work after the first day or two in Shallow Draining, and then determine the amount to be given per acre or chain for the work, and if it should turn out worse in the cutting than I anticipated, to make a reasonable allowance in addition, so that the men may be paid. I may with propriety here mention, that I have seen parties conducting " Shallow Draining" themselves, who have been giving their men 2s. 6d. per acre for their work, whilst mine have been doing the same description of work for 1s. 6d., and have made more at it than those at 2s. 6d.; but this is use, the one party being about to learn their work, under indifferent masters, the other having learnt it in practical working. The prices of work have, I am aware, varied very much throughout the kingdom; and no general system has been adopted, and in some districts, where labour has been plentiful, the work has been done for two-thirds of the amount which it has cost to perform it in others; but with labour, as with an article, it may be bought too cheap. It is not that which costs the least, that wears the best; and so with labour, it is not the cheapest work which is performed in the most proper manner. I have heard parties state, that they could get Shallow Draining done for 8d. and 10d. per chain of twenty-two yards, and so they might, only these same parties should have given us the information that they had ploughed out the furrow twice previously, and that the men only took one draw with the tool—and this they could afford to do. This practice I cannot recommend. I want the whole doing by the tool, and that properly. The lowest price at which you can do it, to get an average of about sixteen inches deep, will be 1s. per acre, without including the tiling; but, as I have before laid down, that twenty inches is sufficiently deep in clay, the price would be, in the ordinary course of Shallow Draining, 1s. 3d. per chain, including everything, and in cases where the land is more open, the scale of prices will run as follows:—

				Per Acre.	
In.		In.		s.	d.
Under 20				1	3
Above 20	and under	24		1	6
„ 24	„	28		1	9
„ 28	„	32		2	0
„ 32	.,	36		2	3

and in some cases 2s. 6d. for thirty-six inches. If it is half hack-work, I allow half as much more, and from that to twice as much, depending upon the strata. In Shallow Draining I have allowed 4s. 6d. and 5s. per chain, and at a depth of from twenty to thirty inches, the men have made in long days barely 2s. 6d. per day, but these are extraordinary instances, and very few of them to be met with. The above will be found to be about the proper scale of payment in all ordinary cases, and includes the tiling as well. Of course in Shallow Draining grass, where the parties wish to have the sod put on the top again, and to leave it as at first, it will require an extra charge, depending upon the mode of replacing the sod. My plan is, with the straight shovel to take off the sod, and lay it by the side, taking the soil with it, and when the drain is filled up with soil to the level of this, the sod is replaced, and the whole will in a short time appear as though it had never been disturbed, and this I vary from 3d. to 6d. per acre, as the parties wish it to be laid. Some persons do not mind how it is laid in again, whether grass or soil upwards, whilst others wish it to be levelled, and of course the more you do at it, the greater the expense.

In the prices for "Deep Draining," the like difficulties will occur, as in the other mode of draining, depending upon the soil. In the ordinary course of cube cutting, the price per yard has been 6d., and this price will be sufficient where the parties have a great quantity of room to work in, including stone, &c., as well as soil; but in "Deep Draining," the case is different, as they are confined to a limited space, and frequently cannot get more than half the soil or strata to be removed out of their way: that this of itself causes the work to be of more value, and whilst the cutting of one chain will not be more than 30s. the next will be £3. and even more, as no more stone, gravel, or rock is to be removed than the actual opening of the drain will admit of. The usual mode of calculation for this work has been in the following manner:— if the first chain will average two feet, it is 2d. per yard, or 1d. per foot deep, per yard run, from one foot to six feet; every additional foot from six feet has been 1½d. to nine feet; and from nine feet to twelve feet, 2¼d.; and so on for every additional three feet to twenty feet or more, being the same proportion or amount. The scale or rate of measure-work in hack-work is nearly the same, taking into account the number of feet of digging at the price above-named, and the hack-work in the following proportions:—*viz.*

Tool-work.			Hack-work.		
Ft.	s.	d.	Ft.	s.	d.
3	0	3	1	0	1½
4	0	4	2	0	3
5	0	5	3	0	4½
6	0	6	4	0	6¼
7	0	8½	5	0	9
8	0	11	6	1	1½
9	1	1¼	7	1	8¼
10	1	3¾	8	2	6

The prices, as stated above, may appear high, but they are only in proportion to the labour performed, and will very often barely pay the men their daily wages, work as hard as they may; and at the same time it should be taken into consideration, that no allowance is made for the earth giving way in the running measures, and if tons come in upon the work, it has all to be removed again, without any additional charge;

and the above also includes the putting in of the tiles, but not the labour of getting the sods, stubble, &c.

In "Bastard Draining," the expense of cutting the drains will depend upon the depth, as if the whole of the work can be done with the draining tool, it will all be digging; but should it turn out to be intersected by veins of gravel, a portion of it will require hacking, and consequently the expense will be greater: the price per chain will vary from 3s. to 12s. and 15s. per chain, according to the average depth of the cutting; and the same prices will apply, as in the ordinary course of "Deep Draining." It is next to impossible to make a bargain with the men, as to the price of the work, as they cannot tell how it may turn out. It may be as advantageous to the master as to the men to make the bargain in the first instance; but if the work is under-let, or not at its fair value to make wages, men are apt to shirk their work, or not perform it properly. The fairest way is to measure it off, and to a practical man, who knows how to value labour, it will be easy to ascertain its worth. The price is not regulated in any case by the width of the drain, as in some it will stand at two feet wide, so that the tiles may be laid in securely, whilst in other places it will not stand at all, and will give way to the extent of four or five feet, as in Deep Draining; thereby causing additional labour to the Drainer in casting. The expense of the cutting for pipe-holes is 6*d*. per cube yard, the same as paid for baring, getting stone, &c. The cost of tiling and sodding will be included in the price per chain, according to the work; and the filling in and casting may be done either by the day or by the chain. If extra labour is required by the tapping or piping, so that two men are occupied in lading out the water, of course an allowance must be made in the work, as it is not reasonable to occupy their time without any additional allowance, and also for paring the sods, if there are any in the field, and laying them ready for putting in. It generally occupies the time of one man, as soon as the drains get to about four feet deep, to hand the tiles, stubble, sods, &c., so as to keep the man at the bottom of the drain constantly employed; and this will apply in all cases below this depth;—this man must be paid by the day.

The expense of boring must, like other things, be in a great measure regulated by the nature and quality of the strata, through which the rods have to pass, as in some instances many yards may be accomplished in a day, whilst in others, perhaps not more than one, and this one attended with more manual labour than all the others put together. I have generally commenced with 1*s*. per yard to five yards, and from thence to ten yards 1*s*. 6*d*, and so in proportion, increasing the amount every five yards, according to the nature of the work, as if the obstructions to be met are not great, the same price may do for ten yards more ; and if they are, the price must be increased in proportion ; so that in all cases the amount paid should be in proportion to the labour performed.

DRAINING TILES.

HAVE BEEN IMPROPERLY MADE, AND BADLY BURNT.—THE DIFFERENCE IN VALUE REGULATED BY THE CARRIAGE AND PRICE OF COALS, ETC.—THE TWEEDDALE COMPANY'S TILES AND HOME-MADE TILES COMPARED.—THE SIZES OF TILES REQUISITE FOR ALL DESCRIPTIONS OF DRAINING.—THE PRICES PAID FOR MAKING AND BURNING THEM.—THE SELLING PRICES WHEN MADE.—GATEWAY TILES, THEIR USE.—AND REMARKS TO THOSE WHO MAY WISH TO HAVE YARDS UPON THEIR OWN ESTATES.

It is not until within the last twenty years, that the manufacture of tiles appears to have been at all considered, or any regard paid either to the size of the opening, the thickness, or the length which they might or ought to be made. True it is, there are exceptions to this rule, as yards are in existence in some parts of the country, but more particularly in Nottinghamshire, where they have studied to make tiles not only suitable in size for draining, but have also selected clay peculiarly adapted for making them; burning them also to that proper degree of hardness which is essentially necessary, before they are fit to be put into the ground.

If the tiles used for "Deep Draining" are not

properly or sufficiently burnt, of what use is it to put them into the ground? for, if they should give way, the drain will be stopped, as a matter of course. I have seen tiles sent from different yards, that have been burnt so badly, that the man laying them in the ground has taken the sides between his thumb and finger, and brought them in two, direct from the centre; and not only have they been improperly burnt, but the shape of them has been such that they were not adapted for resisting the pressure on the top of them. These remarks not only apply to those used for the purpose of "Deep Draining," but also those used for "Shallow Draining." How many thousands and tens of thousands of mis-shapen, badly burnt, and rubbishly-looking tiles, may have been seen led into the fields for use, some of them adapted to suit any description of curve, but few of them for fitting to one another in a straight line. Again, this description of tile, with a hole or opening large enough for a Drainer (whose hands get larger by constant use of the tools) to get his thumb up them, and these are called Draining Tiles; added to which, they have been put into the land at all depths, from six to thirty-six inches, without any regard to the quality of the soil, or without any single thing to uphold them but the subsoil. These are the implements, and this is the practice, which have been adopted in hundreds of instances.

Such a thing as a flat tile for draining, was, until latterly, nearly unknown by half the community, and that would form a sufficient excuse for their not using them. Those who did know them, have given themselves very little more trouble with regard to their use and utility, than to the principles of draining itself.

Even in the present day, at the time I am writing, we have Draining Tiles of all heights and sizes, from the patent Draining Tile made by the " Tweeddale Draining Tile Company," to that made by the least yard in the kingdom, and not only of different sizes, but of all sizes except the proper one, at prices varying as much as 10$s.$ per thousand, in many instances which have come under my notice. This may be accounted for by the variation in the first cost, or price of burning, which latter forms an important item in the charge.

It may not be necessary for me, with the present patents before us, to discuss the amount of the cost or expense of tiles, made for the express purpose of Draining; but the great difference in the expence is the quantity and price of coals required for burning, and the length of time it is requisite, to burn them. There are some kilns where the clay, when made up, will require four days to burn it properly, and others five; therefore, the longer the fire and attendance is required, the expense of the tiles will be increased in proportion, and of course the selling price as well, and consequently the greater the price of an article to drain with, the greater the expense of the draining itself. There is another drawback besides the length of time required in burning, and the cost price of the coals in each place, and that is the carriage of the coals. In some places the coals lay close upon the spot, or near the yard, whilst in others they have to be carried a distance of five or six miles, and in inland places, considerably more. I know of one yard in Nottinghamshire, where the distance is still greater, and those who led the tiles did not receive the leading in money, but a ton of coals for every ton delivered into the yard (which of course they had to lead to their own homes, or where else they pleased,) but even this did not augment the price at the yard, compared with others, which might be named, as they only amounted then to 15$s.$ 6$d.$ per ton, whilst within a few miles further, the yards laying near the Trent, the coals cost as much by water carriage, at all events with the team labour, quite as much. These things are often lost sight of by parties when comparing the relative value of two articles of the same class, only made at different places, and to illustrate this more fully, I will here state the price of a tile, the opening about five inches by six inches, varying perhaps half an inch in the three tiles; at a yard near Southwell, Nottinghamshire, the price of this tile per thousand is £4., the same tile at Bourn, £6.; whilst at Helpstone, near Peterborough, at Mr. Clark's yard, the same tile is £3. 10s. Now either the cost in labour, the coals, the carriage, or the nature of the clay, must form the reason for this extraordinary difference in the value of the article, or else at the place where the largest amount is asked, the demand must be limited, or the profit very great. These are important facts for those having draining to perform, to examine into before they commence; and the quantity to be done, should determine them, if it is best to obtain them in their own neighbourhood, or fetch them from a distance, where a saving can be effected; or manufacturing them upon their own estates.

The Tweeddale Company, who have lengthened their tile to fifteen inches, with an opening of about three inches, as well as their flat tiles, give an addition of one in every four of those made in the home yards, which average only a foot. The cost of these tiles at their yard at Burringham, in Lincolnshire, is 26s. per thousand, and for flats, 15s. per thousand; these are for "Shallow Draining" only, and are very good, as is also their large tile for the outfall, for which they charge 42s. per thousand at their yard, and for a few miles in the neighbourhood, these prices do very well. But if these tiles are to be delivered further inland, they will cost from 3s. to 6s. per ton, according to distance, and one thousand will weigh better than three tons, thus making them from 35s. to 44s. per thousand, to say nothing of the breakage in the vessels; and this will also apply in the same ratio to their large tiles; and not only to this yard, but others, where the same amount of carriage has to be effected. Such being the case, where is the difference between them and those made in our home yards, of the same size in the bore or opening, but about three inches shorter, being only twelve inches long or a little over, and generally sold at 26s. or 28s. per thousand, with an allowance of 1s. in the pound for cash. From being shorter they are less likely to break by carting. The difference is this, that what we gain in length, we lose in the carriage, which makes the patent and home-made tiles equal, or nearly so.

But it is not so much with the price of the tile, as it is with the size of it, that I wish to draw attention. These tiles are made for only one purpose, *viz.*, "Shallow Draining;" and it is quite clear, as I have before shown, that land never could or ever will be drained by this description of draining alone. I shall here give the sizes of tiles requisite for draining, of all descriptions; and if some of my readers should ever happen to go into the neighbourhood of Southwell, or into any of the yards within ten miles or so of that place, they may meet with tiles not only of the proper size, but also of a nature and quality adapted to suit every description of draining, and the tiles in that neighbourhood may be equalled, but cannot be surpassed, by any other yards in the kingdom, either for make, shape, or durability.

I will next come to the size of the tile, and its shape, and in attempting to describe this, I need hardly state, that if the sides of an arch are not carried up perpendicularly, so as to carry the arch itself, it will readily give way; and so it is with tiles. In giving this statement here, I do not act upon my own opinion solely, they were made and used by a man who, in his day, understood more about the drainage of land than almost any other man in this country, and they have since his time been used in the drainage of thousands of acres of both "Shallow and Deep Draining;" and those made and put into the ground nearly thirty years ago, have been examined and found to be as good and sound, and to answer all the purposes of drainage, as effectually as when it was first performed, and as unlikely to perish as they were upon the day when they were first used.

Every man is aware that nothing is lost by having a thing made large enough and good enough, and that, to apply the term of an old friend of mine, it too often happens, "We spoil the ship for a pennyworth of tar." Instead of doing our work effectually in the first instance, we either mar it or have it to do over again, when it might as well have been done properly at the first, and these remarks will equally apply to tiles.

The sizes which I recommend are as follows, *viz.*, seven in number, and the moulds for all of them thirteen inches long :—

	Drain Tiles.			Flat Tiles.	
	Width.	Depth.		Breadth.	Length.
	In.	In.		In.	In.
No. 1	3 by	4	....	7 by	13
,, 2	4 ,,	5	....	8 ,,	13
,, 3	5 ,,	6	....	9 ,,	13
,, 4	6 ,,	7	....	10 ,,	13
,, 5	7 ,,	8			
,, 6	8 ,,	9			
,, 7	9 ,,	10			

The dimensions here given are intended for the clear opening of the tile, that is, within. The thickness of the tiles must be regulated according to the nature of the clay used, as, if it is of good quality, they will not require to be so thick to ensure keeping their form in burning, as if of middling quality.

The above sizes for tiles are both as small and as large as required for all purposes; the two last not so much as the others, unless there is a very great body of water to contend with in the outset, and then not more than from one to two hundred yards, depending upon the length of cutting and the water; but as I have before observed, it is always better to begin with a tile large enough, than one too small, as you may perhaps find that when you have got a considerable distance, you have got more water than your tile will carry, and either have to run up another tile, or carry a branch drain into the main drain, or else, instead of having but one outlet, to have two or more; all this will be prevented by having the tile large enough at first.

With respect to flat tiles, the fourth size flat is as large as required, for, if necessary to use the No. 6 or 7 draining tile, the flat can be used the broad way instead of the long one, and answer the same purpose. A great deal is required in the temper of the clay necessary for making the flats, so as to get them free from cracking both before and after they are burnt. I have known instances in making them, where not more than six out of twelve, or one half, have remained sound after they have been taken out of the kiln; true it is, they have not been entirely spoiled, as they may be used at the outset of a drain, or for flats for smaller drains, but they cannot be used in the deep cuttings.

The cost of making the tiles before-mentioned, I have stated, will depend upon the distance, price, and carriage of coal, &c.; the prices for making and burning, everything being found for the brickmaker, will be as follows:—

	Drain Tile.				Flat Tile.	
	£	s.	d.		s.	d.
No. 1	0	16	6	per 1000	10	0
„ 2	1	1	0	„ do.	12	0
„ 3	1	15	0	„ do.	13	6
„ 4	2	5	0	„ do.	15	0
„ 5	4	0	0	„ do.		
„ 6	6	0	0	„ do.		
„ 7	6	0	0	„ do.		

The above are the prices for making them well, and no allowance is made for loss by the brickmaker; he is only paid for a good saleable article. It may be argued that the above prices are high—they may be done for less, I have no doubt;—and Nos. 3 and 4 have been made in some places for 30s. and 40s., being 5s. less for each; but taking one with the other, the above are the sums which are paid for making the tiles in several counties. The cost of making the gutter brick for conveying water, either open or enclosed, is 4s. 9d. per hundred; and of the ordinary pipe tiles 4d. per yard. The prices are alike for 6 and 7, the risk upon both being nearly equal, and, as before stated, no great quantity of these tiles is required, and the amount charged is not great, compared with the room occupied in the kiln.

The next point is the selling price, and many of those who have made tiles, as before described, have done so principally for their own use; but as there are yards which come within half an inch or so of the above sizes, I shall give the prices at the yard of Mrs. Singleton, near Southwell :—

Drain Tile.

	Ready money price.				Credit price.		
	£	s.	d.		£	s.	d.
No. 1	1	10	0	per 1000	1	11	6
,, 2	2	0	0	,, do.	2	2	0
,, 3	2	10	0	,, do.	2	12	6
,, 4	4	0	0	,, do.	4	4	0
,, 5	5	10	0	,, do.	5	15	6
,, 6	10	0	0	,, do.	10	10	0

and for the large tiles 6d. each, of course making, where a large lot is taken, a difference in the amount, in proportion to the above allowances; and for flat tiles £1. 10s. per thousand. The prices are not increased for a month, or, where parties are taking a quantity regularly, but the credit prices are charged to those who take long credit. The price of the small tile is greater by 4s. than in some yards, but then the article is well made, and larger than what they are in other yards, and are adapted for the purpose of Shallow Draining.

The No. 7 tile is a valuable tile for gateways, as it can be used either by itself or with a course or two of bricks to rest upon, and may be increased to any height required. I have also seen them used one over the other, or with one wrong way upwards and the other on the top, where bricks were scarce, and where a large quantity of water had to pass; they are also particularly valuable for this purpose, as they can be laid without the aid of a bricklayer, being put in by the drainer, and very soon executed; they are more used for this purpose than " Deep Draining."

To those who may be disposed to make their own tiles, I would recommend the use of a pair of metal rollers, for the purpose of grinding the clay, and making it more available for the purposes to which it is applied, and it also removes pieces of stone or other things found in the clay, which very often, after the process of burning, (if not got out,) swell, and fly and crack the tiles, and all this is prevented by the employment of the above means.

I think I have given all the information requisite for attaining, not only the proper sizes, but also the expense of making the tiles. The cost of erecting kilns and sheds upon estates, where required, may be accomplished at an expense of from £100 to £200, depending upon the size of them, and when once erected they are useful for the purposes of making bricks, (even in the neighbourhood of stone,) and their value for building purposes over stone, in all cases which have come under my notice, taking into consideration the materials and the carriage of them, with bricks at £1 8s. per thousand, is in proportion over the stone, as near as the statement I have made of the difference in value of tiles over stone, for the purposes of draining;—and there is the like difference in the facilities or dispatch, in executing the two modes of work.

CONCLUSION.

REMARKS UPON OPEN DYKES, ETC.—THE DRAINAGE OF TOWNS.—AND THE MEANS AFFORDED BY THE ACT OF PARLIAMENT FOR GETTING THE DRAINAGE OF ESTATES EXECUTED.

In the foregoing pages, I have endeavoured to lay down the *true* principles of Draining, and the mode of applying them throughout the kingdom. I may be wrong in some of the statements, but I am satisfied that land can only be effectually drained (at present) by the plans I have proposed. There is one or two points, however, which I have not yet remarked upon, and with which I shall close my subject.

To see and know a good farmer, without making inquiry, but merely from observation, like the old proverb, of "Shew me the company he keeps, and I will tell you the man," I would say, shew me a farm, and after seeing the hedges and ditches, I will tell you the character of the farmer. It is a fact, that these points are, by many farmers, very much lost sight of, and as long as some of them can get the land worked, they do not care either for the borders, the hedges, or the ditches:—the first are allowed to grow a crop of all descriptions of seeds, for stocking a farm, of any quantity or quality but those required;—the second are allowed to run up to a height that hedges were never required or intended to grow, keeping both the sun and the wind from the land; both of which are essential to the growth of the seed, and other produce;—and the third, coming within the scope of my observation and remarks, are allowed to grow and silt up, checking and keeping the water upon the land, instead of carrying it off. It is as important to trim your hedges, and scour your ditches, as it is to drain your land, but more particularly the latter, as, if your drains or ditches are allowed to fill up, of what use is it putting tiles into the land for draining, if your water cannot get off. I would urge upon the attention of my readers, the propriety of always attending to this at the spring and fall of the year; the latter, for properly cleaning and paring off the sides, and clearing out all leaves, &c., so as to render it capable of discharging all the water as quick as possible through the winter, and the former, for removing any deposits which may have accumulated during that period, and to make it available for carrying the water off through the summer. These may appear trifles, but they are necessary and essential for the purposes of good husbandry. There is one evil too with regard to fence ditches; parties making them do not give them width and depth enough, and instead of giving the sides a proper batter or fall, they make them nearly perpendicular, and thus the sides keep constantly giving way, and filling up the bottom. If you cut a dyke, let it be wide and deep enough, to secure it from this. If you are obliged to have a dyke so deep as to be a nuisance, put a tile in, large enough for carrying the water, secure it well, and cover it in. It is not necessary to cut one as large as a canal, to carry as much water as would run through the crown of your hat, and be obliged to keep it open; this can be remedied by the plan I propose.

There is another point, and that is, that the drainage of land has, in a great measure, assimilated to the drainage of towns. In the early stage of places becoming inhabited, but little care was bestowed upon the drainage, and it was no unusual thing, a few years ago, to see towns with large open dykes in them, to carry off the refuse water made by the inhabitants, with, perhaps, one or more tunnels to effect this object; but as the population increased, these supplies increased, and fresh parties kept making additional inlets into these tunnels, and many below the level at which these tunnels were laid; such being the case, it was no wonder that these places became stopped up, and then had to be taken up and relaid at lower levels. This has been the system pursued in many instances, and is continued at the present day. These stoppages of drains in towns have produced the worst consequences, whereas, if they had been made large enough in the first instance, and one rule had been laid down for carrying not only the main drain, but also the inlets, an universal system of drainage would have followed, whilst they are now done not only piecemeal, but are made of all sizes, and very little regard paid to the size of the main drain. The branches have also either not been large enough, or else have been larger than the main drain itself: and not until after a portion of filth and stagnant water has been allowed to accumulate, engendering putrid and other fevers, so as to carry off numbers prematurely, have these matters formed a subject of inquiry, or has any remedy been proposed for removing the evil. Land drainage has been the same. The materials used and the size of them, have not been calculated to carry on the process of draining in the manner which was requisite.

Lastly.—There cannot now be any excuse alleged for the land not being drained. By the Act of the 4th and 5th of Victoria, cap. 55, parties are allowed to borrow the money for the purpose of doing this, upon estates where they have only a life interest, and already a Company has been started for the purpose of aiding and assisting those who require it, not only by taking it off their hands, but finding the means for doing it, and having it repaid by instalments. The only advice I would give connected with this subject is, to see that their plans are such as will secure effectual drainage, and I think the parties who have it done should not be encumbered with the expense, until a year or two has elapsed to prove this. It may be necessary, in many instances, to borrow money to effect the drainage. There is nothing connected with land that has paid for the investment of capital, or ever will pay, so well as draining the soil we cultivate for supplying our wants; and by the adoption of the system here laid down, which is, I trust, effectually calculated to do this, we shall confer a great benefit upon the owners of property, by draining their land not only "thoroughly," but securely and for ever.

FINIS.

PRINTED BY H. JOHNSON, STAMFORD.

— Nº 1. —

Plan of a Field Shallow Drained
in the
LORDSHIP OF SOUTHORPE
in the County of
NORTHAMPTON.

Nºs 1, 2, 3. Shew Drains cut 24ⁿ 6ⁱⁿ through the Shellstone to take the Water lying underneath, being Soakage Water from the adjoining Land.

Nº 4. is a Drain deepened from the Outfall to bring the Water from the Shallow Drains at top into Nº 5, and all these Drains empty into the Pond. Nº 6.

The first five Drains empty into the Outfall Nº 7. and so into the Open Dyke, the rest of the Shallow Drains are on an average 18 in. deep and upon the Stone.

The Soil betwixt a Clay and a Marl, very much interspersed with Shells.

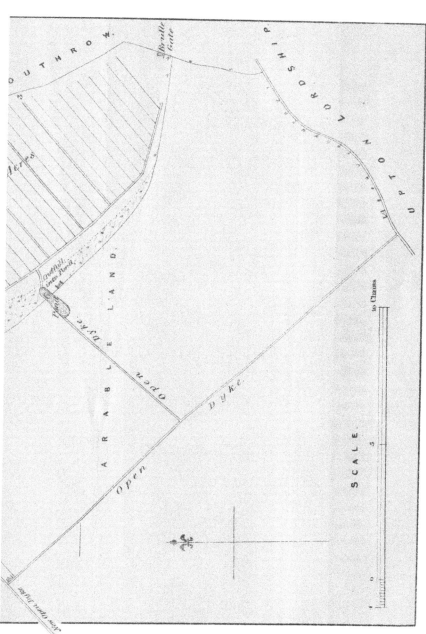

B. Cartwright, Lith. 57, Chancery Lane.

No. 2.

Plan of a Field at
WALCOT
in the County of
NORTHAMPTON.

The Property of Henry Nevile, Esq.

"Bastard Drained" and "Piped" in 1845.

LORDSHIP

BARNACK

Bastard Drain

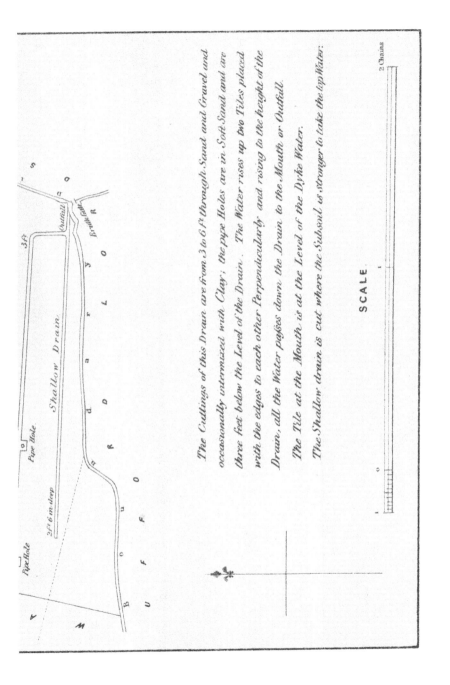

The Cuttings of this Drain are from 3 to 6 Ft through Sand and Gravel and occasionally intermixed with Clay; the pype Holes are in Soft Sand and are three feet below the Level of the Drain. The Water rises up two Tiles placed with the edges to each other Perpendicularly and rising to the height of the Drain, all the Water passes down the Drain to the Mouth or Outfall.

The Tile at the Mouth is at the Level of the Dyke Water.

The Shallow drain is cut where the Subsoil is stronger to take the top Water.

SCALE.

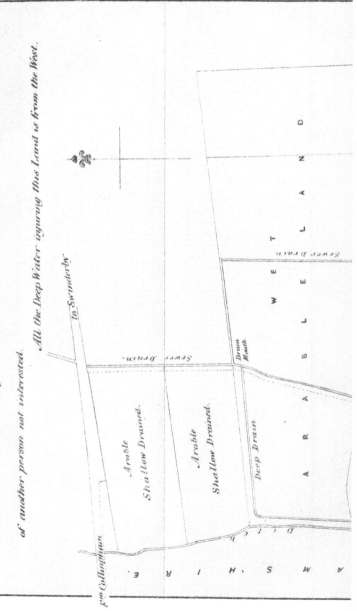

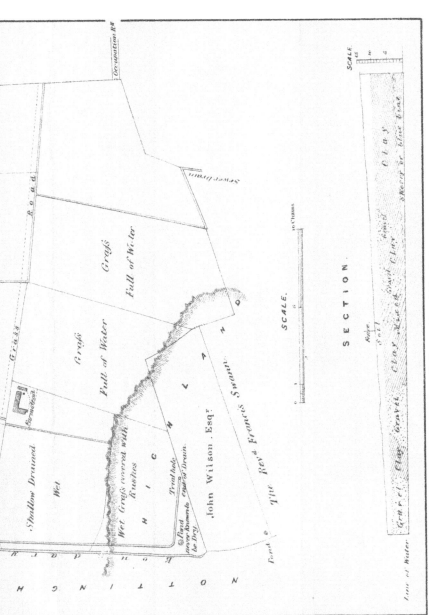

R. Cartwright, Lith. 57. Chancery Lane.

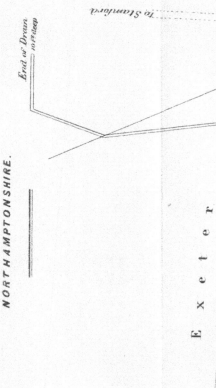

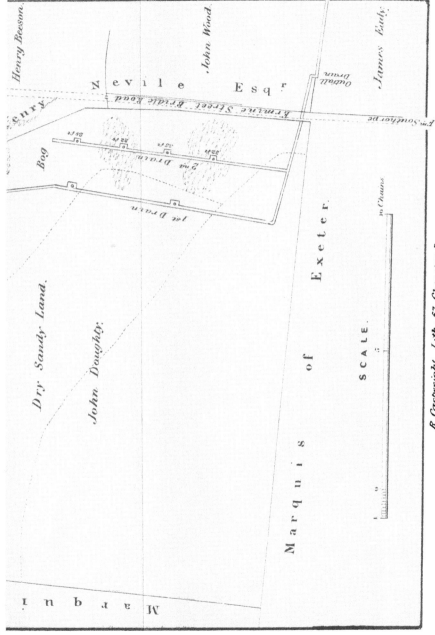